Micro, Mini and Nanodosimetry & Innovative Technologies in Radiation Therapy (MMND & ITRO 2016)

Journal of Physics: Conference Series Volume 777

Tasmania, Australia
26 – 28 January 2016

Editors:

Clive Baldock
Anatoly Rozenfeld
Peter Mecalfe

ISBN: 978-1-5108-3601-3
ISSN: 1742-6588

Printed from e-media with permission by:

Curran Associates, Inc.
57 Morehouse Lane
Red Hook, NY 12571

CURRAN ASSOCIATES INC. proceedings.com

Some format issues inherent in the e-media version may also appear in this print version.

Printed by Curran Associates, Inc. (2017)

For permission requests, please contact the Institute of Physics
at the address below.

Institute of Physics
Dirac House, Temple Back
Bristol BS1 6BE UK

Phone: 44 1 17 929 7481
Fax: 44 1 17 920 0979

techtracking@iop.org

Additional copies of this publication are available from:

Curran Associates, Inc.
57 Morehouse Lane
Red Hook, NY 12571 USA
Phone: 845-758-0400
Fax: 845-758-2633
Email: curran@proceedings.com
Web: www.proceedings.com

Table of contents

Volume 777

Micro-Mini & Nano-Dosimetry & Innovative Technologies in Radiation Therapy (MMND&ITRO2016)

26–28 January 2016, Tasmania, Australia

Accepted papers received: 24 November 2016
Published online: 3 February 2017

Journal of Physics: Conference Series **777** (2016) 011001 doi:10.1088/1742-6596/777/1/011001

Introduction to Mini-Micro-Nano Dosimetry (MMND) 2016

The biennial MMND (formerly MMD) - IPCT workshops, founded in collaboration with Memorial Sloan Kettering Cancer Center (MSKCC) in 2001, has become an important international multidisciplinary forum for the discussion of advanced dosimetric technology for radiation therapy quality assurance (QA) and space science, as well as advanced technologies for prostate cancer treatment.

In more recent years, the interests of participants and the scope of the workshops have extended far beyond prostate cancer treatment alone to include all aspects of radiation therapy, radiation science and technology. We therefore decided to change the name in 2016 to Innovative Technologies in Radiation Oncology (ITRO). MMND ITRO 2016 was held on 26-31 January, 2016 at the beautiful Wrest Point Hotel in Hobart, Tasmania and attracted an outstanding international faculty and nearly 200 delegates from 18 countries (http://mmnditro2016.com/)

The MMND 2016 program continued to cover advanced medical physics aspects of IMRT, IGRT, VMAT, SBRT, MRI LINAC, innovative brachytherapy, and synchrotron MRT. The demand for sophisticated real time and high temporal and spatial resolution (down to the submillimetre scale) dosimetry methods and instrumentation for end–to-end QA for these radiotherapy technologies is increasing. Special attention was paid to the contribution of advanced imaging and the application of nanoscience to the recent improvements in imaging and radiotherapy.

The last decade has seen great progress in charged particle therapy technology which has spread throughout the world and attracted strong current interest in Australia. This demands a better understanding of the fundamental aspects of ion interactions with biological tissue and the relative biological effectiveness (RBE) of protons and heavy ions. The further development of computational and experimental micro-and nano-dosimetry for ions has important application in radiobiology based treatment planning and space radiation hazard prediction. New compact accelerator technologies for the delivery of proton and heavy ion therapy and relevant QA dosimetry instrumentation were an additional focus of MMND 2016.

The ITRO program this year was dedicated to clinical aspects of innovative SBRT for cancer treatment. It represented a unique opportunity to learn from didactic lectures as well as case based discussions with world leaders in the field in the relaxed atmosphere of Hobart.

As well as the outstanding scientific program, MMND ITRO 2016 included an Australian beach BBQ to celebrate Australia Day on the evening of 26th January and an exciting social program on 29th January followed by the conference dinner and great Australian hospitality.

The MMND workshop represents an important next step for improving current cancer treatments with radiation and the development of new radiation based cancer treatments.

MMND 2016 Organising Committee

Prof Anatoly Rozenfeld, PhD
General Chair MMND ITRO 2016
Centre for Medical Radiation Physics
University of Wollongong
Australia

Prof Michael Lerch, PhD
MMND 2016 Co-Chair
Centre for Medical Radiation Physics
University of Wollongong
Australia

Journal of Physics: Conference Series **777** (2016) 011001 doi:10.1088/1742-6596/777/1/011001

Prof Tomas Kron, PhD
MMND 2016 Co-Chair
Department of Physical Sciences
Peter McCallum Cancer Centre
Australia

MMND 2016 Faculty

Dr Chandan Guha, MD

Dr Michael Kraemer, PhD

Dr Malte Ellerbrock, PhD

Dr Chris Beltran, PhD

Prof Mauro Carrara, PhD

 doi:10.1088/1742-6596/777/1/011001

Prof Vladimir Feygelman, PhD

Dr Jay Flanz, PhD

Prof Naruhiro Matsufuji, PhD

Prof Reinhard Schulte, MD

Prof Wolfgang Tomè, PhD

Introduction to Innovative Technology in Radiation Oncology (ITRO) 2016 - *SBRT: Lung, Liver, Brain, Spine and Prostate*

The world is changing with ever increasing rapidity, and the oncologic world even more so. Yet as the writer of Ecclesiastes has observed "The thing that hath been, it is that which shall be; and that which is done is that which shall be done: and there is no new thing under the sun." (Ecclesiastes 1:9) New technologies, such as image guidance, inverse treatment algorithms, exponential increases in computing power and advances in medical imaging technology, have all converged to give birth to stereotactic body radiotherapy or SBRT. There is no question that SBRT is revolutionizing radiation oncology in the 21st century. At the heart of SBRT is the hypofractionation paradigm. In this, very high dose per fraction is given in typically 1-5 fractions with ablative intent, while minimizing target volume margins and normal tissue doses and volumes, allowing extremely high doses of radiation that provide very durable tumor control typically in the range of 90%, while maintaining the risk of severe complications to less than 5-10%. Fractionation has long been the staple of radiation therapy since Coutard first demonstrated that deeply seated head and neck tumors could be cured with fractionation in the

Journal of Physics: Conference Series **777** (2016) 011001 doi:10.1088/1742-6596/777/1/011001

1930's. Fractionation was introduced not out of efforts to improve tumor control, but to limit the effect of radiation on normal tissues, since single session radiation with the limited technology at the time included so much normal tissue that the complications associated with non-fractionated therapy was simply intolerable. SBRT is the technology that has made feasible what we have always known, that high dose per fraction therapy is the most efficient way to kill tumor cells. The marriage of image guidance and highly conformal treatment planning and delivery technology has provided the means to minimize toxicity by limiting the dose to surrounding normal tissue, and creating high dose gradients to concentrate radiation dose within tumors. This is new and wonderful. But the biology of killing tumors, that is, very high dose per fraction radiotherapy to achieve maximal cell kill-- this is nothing new. Indeed, "there is no new thing under the sun."

ITRO 2016 Organising Committee

Dr Josh Yamada, MD
ITRO 2016 Convener
Department of Radiation Oncology
Memorial Sloan Kettering Cancer Centre
USA

Prof Anatoly Rozenfeld, PhD
General Chair MMND ITRO 2016
Centre for Medical Radiation Physics
University of Wollongong
Australia

Dr Michael Jackson, MD
ITRO 2016 Co-Convener
Department of Radiation Oncology
Prince of Wales Hospital
Australia

Dr Joseph Bucci, MD
ITRO 2016 Co-Convener
St George Cancer Care Centre
Australia

ITRO 2016 Faculty

Dr Hirohiko Tsujii, MD

Journal of Physics: Conference Series **777** (2016) 011001 doi:10.1088/1742-6596/777/1/011001

Dr. Josh Yamada, MD

Dr Pat Kupelian, MD

Dr Morten Høyer, MD

Dr Albert Koong, MD

Dr Thomas J Dilling, MD

Dr Carl Rossi, MD

Dr Madhur Garg, MD

Micro-Mini & Nano-Dosimetry and Innovative Technology in Radiation Therapy 2016

International Advisory Committee MMND

- Prof. Stefano Agosteo (Politecnico di Milano, Italy)
- Prof. Clive Baldock (University of Tasmania, Australia)
- Prof. Richard Banati (ANSTO, Australia)
- Prof. Eva Bezak (University of SA, Australia)
- Dr. Elke Brauer-Krisch (ESRF, France)
- A/Prof. Martin Carolan (ICCC, Australia)
- Dr. Stephanie Corde-Tehei (POWH, Australia)
- Dr. Jay Flanz (MGH, USA)
- Dr. Susanna Guatelli (CMRP, Australia)
- Prof. Zdenka Kuncic (University of Sydney, Australia)
- Prof. Sandrine Lacombe (Université Paris-Saclay , France)
- A/Prof. Michael Lerch (CMRP, Australia)
- Prof. Naruhiro Matsufuji (National Institute of Radiological Sciences, Japan)
- Prof. Katia Parodi (Heidelberg Ion Therapy Centre, Germany)
- Dr. Marco Petasecca (CMRP, Australia)
- Dr. Hans Rabus (Physikalisch-Technische Bundesanstalt, PTB, Germany)
- Prof. Anatoly Rozenfeld (CMRP, Australia)
- Dr. Michael Scholz (GSI, Germany)
- Prof. Reinhard Schulte (Loma Linda University, USA)
- Prof. Marco Silari (CERN, Switzerland)
- Prof. Andrey V. Solov'yov (MBN Research Centre, Germany)
- Prof. Taiga Yamaya (National Institute of Radiological Sciences, Japan)
- Prof. Marco Zaider (Memorial Sloan Kettering Cancer Center, USA)

International Advisory Committee ITRO

- Prof. Michael Barton (Ingham Institute, Australia)
- Dr. Joseph Bucci (St. George CCC, Australia)
- Prof. Michael Jackson (POWH, Australia)
- Prof. Paul Keall (University of Sydney, Australia)
- Prof. Patrick Kupelian(UCLA, USA)
- Prof. Tomas Kron (Peter MacCallum Cancer Centre, Australia)
- Prof. Carl Rossi (Scripps Proton Therapy Center, USA
- Prof. Vladimir Feygelman (Moffitt Cancer Center, USA)
- Prof. Morten Hoyer (Aarhus University, Denmark)
- Prof. Peter Metcalfe (CMRP, Australia)
- Prof. Anatoly Rozenfeld (CMRP, Australia)
- Prof. David Thwaites (University of Sydney, Australia)
- Prof. Wolfgang Tomé (Albert Einstein College of Medicine, USA)
- Prof. Hiroko Tsujii (National Institute of Radiological Sciences, Japan)
- Dr. Josh Yamada (Memorial Sloan Kettering Cancer Center, USA)

Local Organizing Committee

- Mr. Saree Alnaghy
- Mr. David Bolst
- Mr. Lachlan Chartier
- Dr. Dean Cutajar
- Mr. Mitchell Duncan
- Ms. Karen Ford
- Dr. Susanna Guatelli

- A/Prof. Michael Lerch
- Dr. Enbang Li
- Prof. Peter Metcalfe
- Mr. Matthew Newall
- Dr. Marco Petasecca
- Dr. Yujin Qi
- Prof. Anatoly Rozenfeld
- Dr. Moeava Tehei
- Dr. Linh Thuy Tran

Conference Proceedings Editors

- Prof. Clive Baldock (University of Tasmania, Australia)
- Prof. Anatoly Rozenfeld (CMRP, Australia)
- Prof. Peter Metcalfe (CMRP, Australia)

MONDAY | 25th January

Pre-Conference Registration
18:00 - 19:00

IOP Conf. Series: Journal of Physics: Conf. Series **777** (2017) 011003 doi:10.1088/1742-6596/777/1/011003

DAY 1 | Tue 26th January

Opening: Micro, Mini and Nanodosimetry Conference 2016

08:00 - 08:20	Coffee and Registration
08:20 - 08:30 **Anatoly Rozenfeld**	Welcome Address
08:30 - 08:55 **Anatoly Rozenfeld**	From MMND IPCT to MMND ITRO : History and overview of CMRP research program

Session 1: Physics of radiation damage and protection in biological tissue

Chairs: Susanna Guatelli | Sandrine Lacombe

08:55 - 09:25 **Roger Martin** Invited Speaker	New DNA-binding ligands that act as topical radioprotectors, mimic endogenous rapid chemical repair
09:25 - 09:45 **Olga Martin** Invited Speaker	Monitoring dna damage response to predict radiotherapy-related normal tissue toxicity
09:45 - 10:00 **Alesia Ivashkevich**	Advances in biological dosimetry
10:00 - 10:30	Morning Tea

MMND & ITRO 2016

Session 2: Nanoparticles and nanodosimetry in radiation therapies

Chairs: Moeava Tehei | Olga Martin

10:30 - 10:55 **Stephen McMahon** Invited Speaker	Gold nanoparticles: Macro- and micro-dosimetric considerations and implications for radiosensitisation
10:55 - 11:20 **Sandrine Lacombe** Invited Speaker	Nanoparticles: Exquisite tools to improve radiation therapy
11:20 - 11:45 **Shivaji Pawar** Invited Speaker	Nanotechnology in Medicine-Physicist Perspective
11:45 - 12:00 **Seung Hoon Yoo** Invited Speaker	Simulation study of dose enhancement in a cell due to nearby carbon and oxygen in particle radiotherapy
12:00 - 12:15 **Moeava Tehei**	Multimodal therapy and multidisciplinary approach: The key for better treatments
12:15 - 12:30 **Bo Li**	Nanodosimetric ionisation cluster size distrubution in view of Heisenberg's Uncertainty Principle
12:30 - 13:30	*Lunch*
13:30 - 13:45 **Susanna Guatelli**	Geant4-DNA: A Monte Carlo package to characterise the effect of high-Z nanoparticles in radiotherapy
13:45 - 14:00 **Yizheng Chen**	DNA damage induced by low-energy photons simulated with nanodosimetry Monte Carlo code - NASIC

Session 3: Dosimetry and microdosimetry in space

Chairs: Anatoly Rozenfeld | Naruhiro Matsufuji

14:00 - 14:25 **Lawrence Pinsky** Invited Speaker	Medipix in space: Past results and future plans
14:25 - 14:50 **Stanislav Pospíšil** Invited Speaker	Use of hybrid Timepix pixel detectors in physics and dosimetry

MMND | DAY 1 | TUE 26th JAN

IOP Conf. Series: Journal of Physics: Conf. Series **777** (2017) 011003 doi:10.1088/1742-6596/777/1/011003

MMND & ITRO 2016

14:50 - 15:05 **Peter Beck** Invited Speaker	A tissue equivalent proportional counter for the assessment of the radiation quality at the International Space Station
15:05 - 15:30	Afternoon Tea

Session 4: Dosimeters in Radiation Therapy

Chairs: Marco Petasecca | Stanislav Pospíšil

15:30 - 15:45 **Filip Vanhavere** Invited Speaker	Feasibility of microdosimetry for carbon beams using mini tissue-equivalent proportional counters
15:45 - 16:00 **Clive Baldock** Invited Speaker	Advancements in gel dosimetry
16:00 - 16:15 **Ikuo Kanno**	Two-dimensional "transXend" detector for energy-resolved computed tomography
16:15 - 16:30 **Francesca Bisello**	Clinical characterization of a monolithic silicon technology for dosimetry in external beam radiotherapy and proton therapy
16:30 - 16:45 **Marco Petasecca**	Development of a silicon diode for skin dosimetry
16:45 - 17:00 **Fumihiko Nishkido**	X-Ray transparent detector for IVR dosimeter using organic photodiodes combined to plastic scintillators

Australia Day: BBQ Dinner

Starts at 18:30

Australia Day is the official National Day of Australia. Celebrated annually on 26 January, it marks the anniversary of the 1788 arrival of the First Fleet of British Ships at Port Jackson, New South Wales, and the raising of the Flag of Great Britain at that site by Governor Arthur Phillip. In present-day Australia, celebrations reflect the diverse society and landscape of the nation, and are marked by community and family events, reflections on Australian history, official community awards, and citizenship ceremonies welcoming new immigrants into the Australian community.

Come together at 6:30pm and celebrate with a BBQ Dinner. Tickets are $95 (+ GST) each and can be purchased on the conference website.

IOP Conf. Series: Journal of Physics: Conf. Series **777** (2017) 011003 doi:10.1088/1742-6596/777/1/011003

DAY 2 | Wed 27th January

Session 1: RBE based treatment planning and QA in Particle Therapy

Chairs: Linh Tran | Peter Beck

08:00 - 08:30	Coffee and Registration
08:30 - 08:55 **Naruhiro Matsufuji** Invited Speaker	Microdosimetric approach to RBE modeling and its implementation in treatment planning and ridge filters design
08:55 - 09:20 **Michael Krämer** Invited Speaker	More Ions for Radiotherapy: About Treatment Planning and Track Simulations
09:20 - 09:45 **Reinhard Schulte** Invited Speaker	Advances in nandosimetric treatment planning for particle beam radiation therapy
09:45 - 10:00 **Malte Ellerbrock** Invited Speaker	Dosimetric quality assurance for scanned ion-beam therapy at HIT
10:00 - 10:30	Morning Tea

IOP Conf. Series: Journal of Physics: Conf. Series **777** (2017) 011003 doi:10.1088/1742-6596/777/1/011003

MMND & ITRO 2016

Session 2: Ion physics and detectors

Chairs: Igor Bray | Koji Noda

10:30 - 10:45 **Alisher Kadyrov** Invited Speaker	Ab initio energy-loss calculations for micro- and nanoscale radiation damage simulations
10:45 - 11:00 **David Hinde** Invited Speaker	Applications of a superconducting solenoidal separator in the experimental investigation of nuclear reactions
11:00 - 11:15 **Mahananda Dasgupta** Invited Speaker	Challenges in describing nuclear reaction outcomes at near-barrier energies
11:15 - 11:30 **Benedikt Bergmann**	Measurement of ionizing energy depositions after fast neutron impact in silicon
11:30 - 11:45 **Linh Tran**	3D silicon microdosimetry and RBE study using ^{12}C ions of different energies

Session 3: Microbeam Radiation Therapy

Chairs: Michael Lerch | Anna Vignati

11:45 - 12:00 **Daniel Häusermann** Invited Speaker	The Australian Synchrotron Imaging and Medical Beamline: New Facilities for in-vivo imaging and radiotherapy
12:00 - 12:15 **Jayde Livingstone**	Dosimetry for spatially fractionated radiotherapy at the Australian synchrotron
12:15 - 12:30 **Michael Lerch**	Innovative semiconductor dosimetry in microbeam radiation therapy
12:30 - 13:30	Lunch

Session 4: Quality Assurance in Hadron Therapy

Chairs: Malte Ellerbrock | Alisher Kadyrov

13:30 - 14:00 **Francesco Romano** Invited Speaker	Status of ELIMED the medical and multidisciplinary beamlines at ELI-beamline
14:00 - 14:15 **Matthias Würl**	Experimental studies with two novel silicon detectors for the development of time-of-flight spectrometry of laser-accelerated proton beams

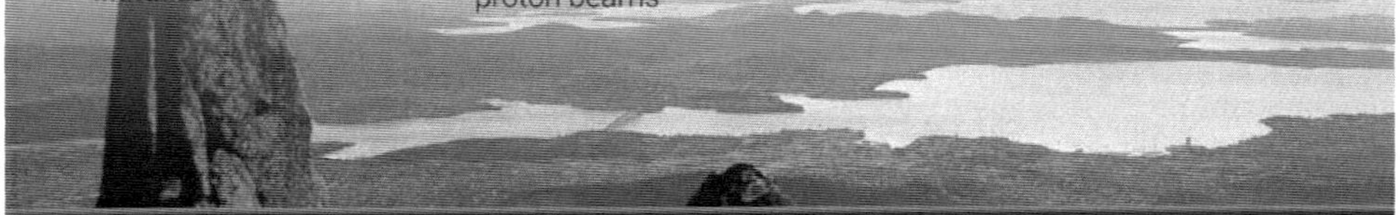

MMND | DAY 2 | WED 27th JAN

IOP Conf. Series: Journal of Physics: Conf. Series **777** (2017) 011003 doi:10.1088/1742-6596/777/1/011003

MMND & ITRO 2016

14:15 - 14:30 **Anna Vignati** Invited Speaker	Fast dose analysis of movement effects during treatments with scanned proton and carbon-ion beams
14:30 - 14:45 **Wenjing Chen**	Treatment verification with off-line PET imaging in particle therapy and its clinical use at HIT
14:45 - 15:00 **Michele Togno**	Characterization of novel solutions for daily QA and patient plan verification in pencil beam proton beams
15:00 - 15:30	Afternoon Tea (with poster session)

Session 5: Advancing X-ray radiotherapy

Chairs: Anna Ralston | Tomas Kron

15:30 - 15:45 **Mauro Carrara** Invited Speaker	Breathing induced prostate motion: Development of a patient's specific model based on Calypso transponders analysis
15:45 - 16:00 **Peter Metcalfe** Invited Speaker	Introducing Dynamic Dosimaging: Potential applications to MRI-linac
16:00 - 16:15 **Jarrad Begg**	The Australian MRI-Linac Program: Measuring profiles and PDD in a horizontal beam
16:15 - 16:30 **Justin Turpin**	The Elkta MRI-Linac
16:30 - 16:45 **Michael Jameson**	Investigating the relationship between treated gross tumour volume (GTV) and metabolic tumour volume (MTV) for non-small cell lung cancer patients
16:45 - 17:00 **Meng Cai**	Linearization of film dosimetry system and reference dosimetry for SBRT quality assurance
17:00 - 17:15 **Anna Ralston**	The current status of small field output measurements using diamond and silicon diode detectors
17:15 - 17:30 **Zoë Moutrie**	Commissioning and clinical implementation of a comerical intrafractional kilovoltage integrated imaging system for stereotactic radiation therapy treatments

MMND | DAY 2 | WED 27th JAN

IOP Conf. Series: Journal of Physics: Conf. Series **777** (2017) 011003 doi:10.1088/1742-6596/777/1/011003

DAY 3 | Thur 28th January

Session 1: Proton and HIT Therapy 1

Chairs: Anatoly Rozenfeld | Carl Rossi

08:00 - 08:30	Coffee and Registration
08:30 - 09:05 **Masumi Umezawa** Invited Speaker	Advanced technology based on scanning irradiation for particle therapy system
09:05 - 09:25 **Koji Noda** Invited Speaker	Development of heavy-ion radiotherapy facility in Japan
09:25 - 09:45 **Taiga Yamaya** Invited Speaker	OPENPET: A novel open-type PET system for 3D dose verification in particle therapy
09:45 - 10:00 **Malte Ellerbrock** Invited Speaker	Clinical operation of the Heidelberg Ion-Beam Therapy Center
10:00 - 10:30	Morning Tea

IOP Conf. Series: Journal of Physics: Conf. Series **777** (2017) 011003 doi:10.1088/1742-6596/777/1/011003

MMND & ITRO 2016

Session 2: Proton and HIT Therapy 2

Chairs: Michael Jackson | Hirohiko Tsuji

10:30 - 10:55 **Jay Flanz** Invited Speaker	Current and future technical challenges in particle therapy
10:55 - 11:15 **Damien Bertrand** Invited Speaker	Protect, enhance and save more lives with proton therapy
11:15 - 11:35 **Carl Rossi** Invited Speaker	The Scripps Proton Therapy Facility
11:35 - 11:50 **Morten Høyer** Invited Speaker	Establishment of the Danish Center For Particle Therapy
11:50 - 12:05 **Reinhard Schulte** Invited Speaker	Advances in proton imaging for treatment planning, verification, and adaptation in particle beam radiation therapy
12:05 - 12:30 **Chandan Guha** Invited Speaker	Radioprotectors in Radiation Therapy
12:30 - 13:30	Lunch

Session 3: Clinical/SBRT 1

Chairs: Peter Metcalfe | Wolfgang Tome

13:30 - 13:55 **Vladimir Feygelman** Invited Speaker	Measurement and calculation of 4D dose
13:55 - 14:20 **Wolfgang Tomé** Invited Speaker	An overview of current and future methods of image guided SBRT
14:20 - 14:45 **Paul Keall** Invited Speaker	Innovative technologies in radiation oncology: One city building two new cancer radiotherapy systems
14:45 - 15:10 **Tomas Kron** Invited Speaker	Patient specific quality control for SABR: One phantom does not fit all
15:10 - 15:30	Afternoon Tea

MMND | DAY 3 | THUR 27th JAN

IOP Conf. Series: Journal of Physics: Conf. Series **777** (2017) 011003 doi:10.1088/1742-6596/777/1/011003

MMND & ITRO 2016

Session 4: Clinical/SBRT 2

Chairs: Paul Keall | Vladimir Feygelman

15:30 - 15:55 **Chris Beltran** Invited Speaker	Clinical implementation of a proton therapy dose verification system based on GPU accelerated Monte Carlo calculations
15:55 - 16:10 **Nicholas Hardcastle**	The LIGHT SABR Clinical Trial: Tumour tracking radiotherapy for superior lung cancer outcomes
16:10 - 16:25 **Johannes Thoelking**	Patient-specific dose verification for 6MV flattening filter free (FFF) beams based on transmission measurements
16:25 - 16:40 **Vaughan Moutrie**	Out of field dose during Gamma Knife treatment: A paediatric case study
16:40 - 16:55 **Sankar Arumugam**	Clinical implementation of SeedTracker for prostate SBRT: Experience and gains so far
16:55 - 17:10 **Shankar Siva**	Effect of stereotactic ablation on normal kidney function – Radiobiological modelling of a dose / response relationship
17:10 - 17:25 **Kathryn Clarke**	Is it necessary to consider optic nerve movement when treating the optic nerve with stereotactic radiosurgery?
17:25 - 17:40 **Michael Jackson** Invited Speaker	Enhancing cancer treatments with oncothermia

MMND | DAY 3 | THUR 27th JAN

MMND & ITRO 2016

DAY 4 | Fri 29th January

Social Day: MONA Muesum

The Museum of Old and New Art (MONA) is an art museum located within the Moorilla winery on the Berriedale peninsula in Hobart, Tasmania, Australia. It is the largest privately funded museum in Australia, known for its playful and through-provoking exhibitions.

As this is not an organised event, you can make your way to the museum on a ferry from the Mona Ferry Terminal, just 3km away from the Wrest Point. Pricing for the ferry ride and museum admission can be found on the museum website. Don't forget to get back to the Wrest Point by 5:30pm for communication with industry, followed by evening drinks and a gala dinner.

Gala Dinner: Wrest Point

Pre-dinner drinks from 18:15

Enjoy pre-dinner drinks with industry from 18:15 - 18:55 in the foyer, sponsored by Hitachi. Then come together at 19:00 in the Wrest Point Derwent Room for a gala dinner. Cost is included in the conference registration (MMND, ITRO or combined). Additional tickets for guests can be purchased from the conference website.

Transportation

A taxi is $10.00-15.00 AUD from Wrest Point to the Salamanca, where both the MONA Ferry Terminal and the Drunken Admiral Restaurant are located. This is generally the easiest option as the Wrest Point has a taxi rank outside the front doors, and cheapest if split with multiple people.

Alternatively, a standard bus fare is $3.10. Timetables are available at the Wrest Point Guest Services desk, who can also assist in giving directions to the bus stop.

Finally, if you enjoy a walk along the waterfront through Battery Point from Wrest Point to Salamanca, this takes approximately 30-40 minutes.

IOP Conf. Series: Journal of Physics: Conf. Series **777** (2017) 011003 doi:10.1088/1742-6596/777/1/011003

DAY 1 | Sat 30th January

ITRO 2016 Introduction

Anatoly Rozenfeld & Josh Yamada

08:00 - 08:05

Session 1: SBRT CNS (brain and spine)

Chairs: Michael Jackson | Carl Rossi

08:05 - 08:30 **David Currow** Invited Speaker	Cancer treatment in NSW
08:30 - 09:15 **Josh Yamada** Invited Speaker	CNS lecture
09:15 - 09:30 **Josh Yamada** Invited Speaker	CNS lecture questions
09:30 - 10:00 **Josh Yamada** **Eric Hau**	CNS case presentation
10:00 - 10:15	Break (with industry)

IOP Conf. Series: Journal of Physics: Conf. Series **777** (2017) 011003 doi:10.1088/1742-6596/777/1/011003

MMND & ITRO 2016

Session 2: SBRT GI (liver and pancreas)

Chairs: Marketa Skala | Patrick Kupelian

10:15 - 11:00 **Albert Koong** Invited Speaker	GI lecture
11:00 - 11:15 **Albert Koong** Invited Speaker	GI lecture questions
11:15 - 11:45 **Albert Koong** **Andrew Kneebone**	GI case presentation
11:45 - 12:00	Break (with industry)

Session 3: SBRT thoracic (lung)

Chairs: Verity Ahen | Madhur Garg

12:00 - 12:45 **Tom Dilling** Invited Speaker	Lung lecture
12:45 - 13:00 **Tom Dilling** Invited Speaker	Lung lecture questions
13:00 - 13:30 **Tom Dilling** **Marketa Skala**	Lung case presentation
13:30 - 14:30	Lunch

Session 4: SBRT head and neck

Chairs: Shankar Siva | Albert Koong

14:30 - 15:15 **Madhur Garg** Invited Speaker	Head and neck lecture
15:15 - 15:30 **Madhur Garg** Invited Speaker	Head and neck lecture questions
15:30 - 16:00 **Madhur Garg** **Andrew Miller**	Head and neck case presentation

ITRO | DAY 1 | SAT 30th JAN

IOP Conf. Series: Journal of Physics: Conf. Series **777** (2017) 011003 doi:10.1088/1742-6596/777/1/011003

MMND & ITRO 2016

Session 5 Immunotherapy

Chairs: Morten Høyer | Joseph Bucci

16:00 - 16:20 **Chandan Guha** Invited Speaker	Immunotherapy lecture
16:20 - 16:30 **Chandan Guha** Invited Speaker	Immunotherapy lecture questions
16:30 - 19:00	Free time

Social Dinner: The Drunken Admiral

Starts at 19:00

The Drunken Admiral has been operating as a family run restaurant since 1979. The diner provides a truly unique experience, offering the very freshest Tasmanian seafood and meals set amongst a vibrant maritime atmosphere. Being one of Hobarts most iconic restaurants, it's popular with both locals and tourists alike.

Get together with the MMND ITRO Faculty at the Drunken Admiral at 7pm. Tickets can be purchased for $85, which includes a set menu + drinks. There are limited tickets available due to the size of the restaurant - first in, best dressed! See conference website for details.

Transportation

A taxi is $10.00-15.00 AUD from Wrest Point to the Salamanca, where both the MONA Ferry Terminal and the Drunken Admiral Restaurant are located. This is generally the easiest option as the Wrest Point has a taxi rank outside the front doors, and cheapest if split with multiple people.

Alternatively, a standard bus fare is $3.10. Timetables are available at the Wrest Point Guest Services desk, who can also assist in giving directions to the bus stop.

Finally, if you enjoy a walk along the waterfront through Battery Point from Wrest Point to Salamanca, this takes approximately 30-40 minutes.

IOP Conf. Series: Journal of Physics: Conf. Series **777** (2017) 011003 doi:10.1088/1742-6596/777/1/011003

DAY 2 | Sun 31th January

Session 1: SBRT GU (prostate)

Chairs: Andrew Kneebone | Tom Dilling

08:00 - 08:45 **Patrick Kupelian** Invited Speaker	GU lecture 1
08:45 - 09:00 **Patrick Kupelian** Invited Speaker	GU lecture 1 questions
09:00 - 09:45 **Morten Høyer** Invited Speaker	GU lecture 2
09:45 - 10:00 **Morten Høyer** Invited Speaker	GU lecture 2 questions
10:00 - 10:30 **Morten Høyer** **Patrick Kupelian** **Joseph Bucci**	GU case presentation
10:30 - 10:45	Break (with industry)

IOP Conf. Series: Journal of Physics: Conf. Series **777** (2017) 011003 doi:10.1088/1742-6596/777/1/011003

MMND & ITRO 2016

Session 2: Particles (protons and carbon)

Chairs: Michael Jackson | Chandan Guha

10:45 - 11:15 **Carl Rossi** Invited Speaker	Advancements in proton therapy
11:15 - 11:45 **Hirohiko Tsujii** Invited Speaker	Carbon ions
11:45 - 12:00 **Carl Rossi** **Hirohiko Tsujii**	Particle therapy questions
12:00 - 12:30	Lunch (with industry)

Session 3: Special topics

Chairs: Tomas Kron | Mauro Carrara

12:30 - 12:40 **Anatoly Rozenfeld**	Innovative quality assurance for SBRT
12:40 - 13:00 **Paul Keall** Invited Speaker	Bench-to-bedside implementation of real-time image guidance and adaptation in prostate and lung SBRT
13:00 - 13:20 **Vladimir Feygelman** Invited Speaker	Effect of respirtory motion on target dose
13:20 - 13:40 **Wolfgang Tomé** Invited Speaker	Spine SBRT using VMAT
13:40 - 14:00 **Anatoly Rosenfeld** **Paul Keall** **Vladimir Feygelman** **Wolfgang Tomé**	Questions

Closing and future meetings

Anatoly Rozenfeld & Josh Yamada

14:00 - 14:10

ITRO | DAY 2 | SUN 31th JAN

Peer review statement

All papers published in this volume of *Journal of Physics: Conference Series* have been peer reviewed through processes administered by the proceedings Editors. Reviews were conducted by expert referees to the professional and scientific standards expected of a proceedings journal published by IOP Publishing.

IOP Conf. Series: Journal of Physics: Conf. Series **777** (2017) 012001 doi:10.1088/1742-6596/777/1/012001

History of International Workshop on Mini-Micro- and Nano-Dosimetry (MMND) and Innovation Technologies in Radiation Oncology (ITRO)

Anatoly B. Rosenfeld[1], Marco Zaider[2], Josh Yamada[2], Michael J.Zelefsky[2]
[1]Centre for Medical Radiation Physics, University of Wollongong, Wollongong NSW 2522, Australia
[2]Memorial Sloan Kettering Cancer Centre, NY, USA

Email: anatoly@uow.edu.au

Abstract. The biannual MMND (former MMD) - IPCT workshops was founded in collaboration between the Centre for Medical Radiation Physics, University of Wollongong and the Memorial Sloan Kettering Cancer Center (MSKCC) in 2001 and has become an important international multidisciplinary forum for the discussion of advanced quality assurance (QA) dosimetry technology for radiation therapy and space science, as well as advanced technologies for clinical cancer treatment.

1. Historical overview

Microdosimetry is the systematic study and quantification of the *spatial* and *temporal* distribution of absorbed energy in irradiated matter. It deals with *stochastic* energy deposition. Microdosimetry is important for characterization of mixed radiation fields for prediction of its biological effect in humans and failure in electronic devices [1].

The idea of the creation of an international forum for discussion on the development of new techniques in microdosimetry and its applications was originated by Prof Anatoly Rosenfeld PhD, of the Centre for Medical Radiation Physics (CMRP) of the University of Wollongong, Australia and Prof Marco Zaider PhD, of the Memorial Sloan Kettering Cancer Centre (MSKCC), USA due to a strong move in the development of a new generation of solid state microdosimetry instruments for space/aviation, radiation protection and radiation therapy. The forum originated at CMRP in 1995 and further boosted by a National Health and Medical Research Council (NHMRC) Project Grant - *Development of Silicon Detectors for Microdosimetry* in 1996. An outcome of this project was the first Silicon on Insulator (SOI) microdosimeter which has demonstrated advancement for microdosimetry in hadron therapy [2].

To address this direction of fundamental and translational research CMRP in partnership with MSKCC founded the Mini-Micro-Dosimetry (MMD) International Workshop. The first inaugural MMD was held from the 16-20th December, 2001 in the Medina Grand Hotel, Sydney, Australia. MMD 2001 highlighted the strong interest in the development and applications of semiconductor radiation detectors, not only for microdosimetry, but also for semiconductor mini-detectors for quality assurance (QA) in radiation therapy. Since 2001 the MMD Workshop has become an important biennial international multidisciplinary forum for the discussion of advanced QA dosimetry technology for radiation therapy including microdosimetry for hadron therapy and space science.

IOP Conf. Series: Journal of Physics: Conf. Series **777** (2017) 012001 doi:10.1088/1742-6596/777/1/012001

The second MMD workshop was held in conjunction with the World Congress on Medical Physics and Biomedical Engineering from the 24-29th of August, 2003 at the Sydney International Convention Centre. The MMD 2003 workshop included interesting discussions led by invited international speakers on ion radiobiology by Dr Mauro Belli PhD of the Instituto Superiore di Santia, Rome; on the cancer risk of space radiation by Prof John Dicello PhD of Johns Hopkins University; on microdosimetry by Prof Roberto Cherubini PhD of INFN; challenges in radiobiology and dosimetry in synchrotron Microbeam Radiation Therapy (MRT) by Dr Alberto Bravin PhD and Dr Elke Braeuer Krisch of ESRF; on proton therapy by Prof Reinhard Schulte MD of LLUMC; on fast neutron therapy by Dr Mark Yudelev PhD of Karmanos Cancer Institute. All demonstrated the importance of solid state microdosimetry in radiation therapy and space.

In 2005 CMRP and MSKCC decided to extend their collaboration in prostate cancer treatment research by founding the International Workshop on Prostate Cancer Treatment (IPCT) with Prof Anatoly Rosenfeld PhD (CMRP) and Prof Michael Zelefsky MD and Prof Marco Zaider PhD (MSKCC) as Foundation Chairs. This first workshop was held in Sydney in collaboration with St Vincent's Hospital and was co-chaired by Prof Phillip Striker MD of St Vincent's Private Hospital.

Since 2005 both the MMD and IPCT workshops have been run together and known as the MMD-IPCT International Workshops. MMD-IPCT 2005 was held from 5-8 December, 2005 at the University of Wollongong, and became a large multidisciplinary forum which brought together radiation detection scientists, medical physicists, radiobiologists and radiation oncologists to cover many aspects of dosimetry and microdosimetry detectors for QA in radiation therapy and space and advancements in prostate cancer treatment with external X-ray and proton beams and brachytherapy. The invited speakers Dr Josh Yamada MD (MSKCC), Prof Carl Rossi MD (LLUMC), Dr Katja Langen PhD (MD Anderson Cancer Center, Orlando) led fascinating discussions on the comparison of radiation oncology modalities for prostate cancer treatment. Since this workshop, Dr Josh Yamada became Convener of IPCT workshops. Many international dosimetry experts joined MMD-IPCT 2005 including Prof Pavel Olko PhD of the Krakow Institute for Nuclear Physics, Dr Alex Romanyukha of the US Naval Dosimetry Center and Dr Henry Tang of IBM among many others.

MMND-ITRO 2008 was held at the University of Wollongong. The feature of MMD 2008 Workshop was a special session on *Microdosimetry for Space Applications* supported by the National Space Biomedical Research Institute (NSBRI), USA in recognition of the developed microdosimrtry instrument, MIDN for astronaut's personal dosimetry based on the CMRP SOI microdosimetry detector in collaboration with US Naval Academy (USNA), NASA and MSKCC which was supported by a joint NSBRI grant. Invited speakers were Prof John Dicello PhD and Prof Vince Pisacane PhD (USNA).

In 2008 a new topic was introduced at MMD - experiments in nanodosimetry and associated Monte Carlo radiation transport simulations. Invited speakers on nanodosimetry simulations were Dr Sébastien Incerti PhD of CENBG, Bordeaux, and Dr Elisabetta Gargioni PhD of University Medical Center, Hamburg; Dr Henry Tang PhD of IBM; and on experimental nanodosimetry, Prof Reinhard Schulte MD of LLUMC. Due to strong interest in nanodosimetry, along with microdosimetry, it was decided to rename the MMD workshop to Mini-Micro-Nano Dosimetry (MMND). The IPCT 2008 workshop was co-chaired by Dr Josh Yamada MD of MSKCC and it became a forum where invited speakers, Dr James Morris and Prof Carl Rossi MD of LLUMC contributed to interesting discussions on the comparison of prostate cancer treatment with proton therapy versus permanent seed implant brachytherapy and HDR brachytherapy.

In 2010 the MMND-IPCT Workshop was held in conjunction with the Solid State Dosimetry (SSD 16) conference from the 19-24th of September, 2010 at the Sofitel Wentworth Hotel in Sydney. Special guest was US NASA astronaut Dr Leroy Chao PhD who presented a talk on importance of monitoring of the radiation environment during human space missions and the role of microdosimetry for personal radiation protection of astronauts.

MMND-IPCT 2012 was held at the University of Wollongong and had a strong emphasis on nanodosimetry and its relation to microdosimetry and the role of nanoparticles in dose enhancement in radiation therapy as presented by invited speaker Prof Claude Lesech of Institute des Sciences Moléculaires d'Orsay. Invited speakers, Prof Larry Pinsky PhD of the University of Houston, and Prof Stanislav Pospisil PhD of the Institute of Experimental and Applied Physics, Prague, led discussions on MEDIPIX detectors with their application for space dosimetry on a board of ISS, and neutron and fragments detection in heavy ion therapy. A joint session with IEEE NPSS on innovative scintillating detectors and applications for advanced SPECT and PET-MRI imaging was presented by invited speakers Prof Katia Parodi PhD of LMU, Munich, Dr Paul Lecoq PhD of CERN, Prof Tom Lewellen PhD of University of Washington and Prof Benjamin Tsui of Johns Hopkins University. Alumni of CMRP made a strong contribution to MMND-IPCT 2012 including Dr Benjamin Clasie PhD of F Burr Proton Therapy Centre, MGH, Dr Andrew Wroe PhD of LLUMC in proton therapy technology and Dr Peter Bradley PhD of Zarlink Semiconductor, USA in wireless implantable sensors technology with possible applications for in vivo radiation dosimetry.

At ITRO 2012 we re-introduced again, since the inauguration IPCT 2005, the case studies in prostate cancer treatments led by Dr Josh Yamada MD with important contribution from invited speakers on proton therapy by Prof Carl Rossi MD and on C-12 ion therapy by Dr Hiroshi Tsuji MD of NIRS followed by discussion on the comparison of clinical outcomes of patient treated by protons and C-12 ions versus IMRT and brachytherapy.

The MMND-IPCT 2014 was held from 20-25 October in Port Douglas Sheraton Mirage Hotel and was a great success due to the very interesting scientific and social programs and an outstanding international faculty (http://mmnd-ipct.com). A special session was devoted to the outcome of the ARDENT project (http://ardent.web.cern.ch/) on the development of innovative detectors for QA dosimetry, solid state microdosimetry in radiation therapy and space. The workshops were attended by more than 150 medical physicists, radiation oncologists, radiation detector experts, radiobiologists and radiation space scientists from 15 countries.

Each biennial MMND IPCT workshop was a giant step in the demonstration of innovative technologies that led to progress in solid state microdosimetry, as recently reviewed [3], and its growing applications in proton and heavy ion radiation therapy and space radiation dosimetry. The workshops demonstrated the importance in collaboration and synergy of radiation detectors and dosimetry experts, radiobiologists, medical physicists and physicians to address challenges in contemporary radiation therapy and new avenues. Confirmation of that was demonstrated in the recent outstanding MMND–ITRO 2016 workshop (http://mmnditro2016.com/).

2. References

[1] Rossi H H and Zaider M 1996 Microdosimetry and its applications, Springer

[2] Bradley P D, Rosenfeld A B and Zaider M 2001 Solid State Microdosimetry. *Nucl. Instrum. Methods Phys. Res. B.* **184** 135-57

[3] Rosenfeld A B 2016 Novel detectors for silicon based microdosimetry, their concepts and applications, *Nucl. Instr. Meth. Phys. Res.* A **809** 156-70

IOP Conf. Series: Journal of Physics: Conf. Series **777** (2017) 012001 doi:10.1088/1742-6596/777/1/012001

3. Memories

Figure 1. MMD-IPCT 2003: John Dicello, Anatoly Rozenfeld (front right) with group of invited speakers including Roberto Cherubini in centre of first row. First row: Michael Lerch, Mark Yudelev, Alesandro Campa, Alberto Bravin. Second row: Elke Braeuer-Krisch, Stephanie Corde, Roberto Cirio (and others).

Figure 2. Michael Zelefsky and Anatoly Rozenfeld (left). Online technology transfer in LDR Brachytherapy by Dr. Zelefsky, MD at St. Vincent's Private Hospital during inaugural IPCT 2005.

Figure 3. MMND –IPCT 2005

Figure 4. MMND-IPCT 2008

Figure 5. MMND-IPCT 2012, University of Wollongong. From the left to the right: Claude Lesech, Reinhard Schulte, Larry Pinsky, Anatoly Rozenfeld, Katia Parodi, Paul Lecoq, Benjamin Tsui and Tom Lewellen.

Figure 6. MMND-IPCT 2014, Port Douglas, Sheraton Mirage. In a centre Mack Roach, Josh Yamada, Anatoly Rozenfeld.

Atlas-based segmentation technique incorporating inter-observer delineation uncertainty for whole breast

L R Bell[1,2], J A Dowling[3], E M Pogson[1,2,5], P Metcalfe[1,2], L Holloway[1,2,4,5]
[1]Centre for Medical Radiation Physics, University of Wollongong, Australia
[2]Liverpool & Macarthur Cancer Therapy Centres & Ingham Institute, Liverpool, Australia
[3]Australian e-Health Research Centre, CSIRO, Queensland, Australia
[4]SWSCS, University of New South Wales, Australia
[5]Institute of Medical Physics, University of Sydney, Australia

E-mail: lb998@uowmail.edu.au

Abstract. Accurate, efficient auto-segmentation methods are essential for the clinical efficacy of adaptive radiotherapy delivered with highly conformal techniques. Current atlas based auto-segmentation techniques are adequate in this respect, however fail to account for inter-observer variation. An atlas-based segmentation method that incorporates inter-observer variation is proposed. This method is validated for a whole breast radiotherapy cohort containing 28 CT datasets with CTVs delineated by eight observers. To optimise atlas accuracy, the cohort was divided into categories by mean body mass index and laterality, with atlas' generated for each in a leave-one-out approach. Observer CTVs were merged and thresholded to generate an auto-segmentation model representing both inter-observer and inter-patient differences. For each category, the atlas was registered to the left-out dataset to enable propagation of the auto-segmentation from atlas space. Auto-segmentation time was recorded. The segmentation was compared to the gold-standard contour using the dice similarity coefficient (DSC) and mean absolute surface distance (MASD). Comparison with the smallest and largest CTV was also made. This atlas-based auto-segmentation method incorporating inter-observer variation was shown to be efficient (<4min) and accurate for whole breast radiotherapy, with good agreement (DSC>0.7, MASD <9.3mm) between the auto-segmented contours and CTV volumes.

1. Introduction

Internal anatomical changes during the course of radiotherapy limit treatment accuracy, as current treatments are planned on static images obtained prior to delivery [1]. Adaptive radiotherapy is an increasingly investigated approach that aims to combat these anatomical changes by re-imaging and re-planning at multiple time points throughout the treatment course. However, this process has large workflow implications due to the time required to achieve this whilst maintaining plan quality.

Target delineation is the bottleneck of the planning process taking on average 18.6 min per patient for breast radiotherapy [2], and housing one of the largest uncertainties affecting modern radiotherapy accuracy [3]. Segmentation techniques, boasting significant decreases in delineation times as compared to manual contouring, have been proposed as the solution to this limitation with several techniques implemented with impressive results [4]. Atlas-based segmentation techniques in particular are suited for adaptive radiotherapy purposes as they are derived from real patient anatomy and verified clinician contours. However, current techniques generally fail to incorporate inter-observer variation, with single clinician contours frequently taken to be the gold standard delineation. With the largest uncertainty in modern radiotherapy treatment occurring during the target delineation process, it is essential to account for delineation uncertainty when devising an adaptive radiotherapy process.

This atlas based segmentation technique improves on the accuracy of existing techniques by incorporating inter-observer variation into the segmentation process, whilst maintaining efficacy.

2. Method

This segmentation technique, although developed and validated for a whole breast, is applicable to any target volume definition in which the contours are defined based on anatomy. The general process involves generating atlas, registering the atlas to a new patient to obtain good correspondence between equivalent regions, and mapping the atlas segmentation to the new image space.

2.1. Technique development

The atlas generation method, outlined in figure 1, has been utilised previously in medical imaging applications [5]. An initial dataset representative of the patient cohort was chosen with remaining datasets in the cohort registered to the initial dataset using rigid and affine transformation [6]. The results were averaged and used as the subject for the next iteration in which rigid, affine registration and diffeomorphic demons non-rigid registration [7] was used. Three iterations were performed to produce the final atlas.

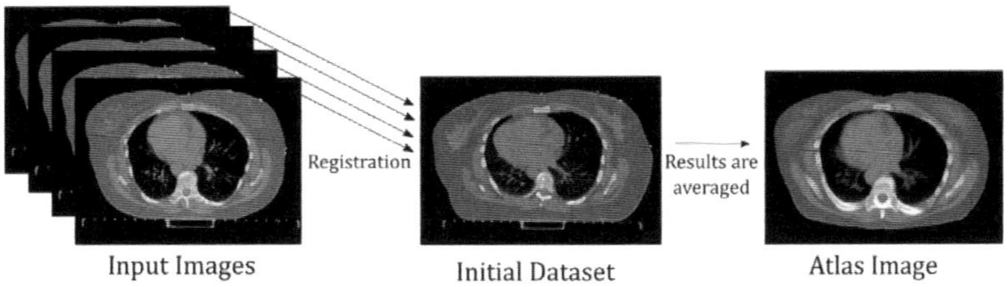

Figure 1. Method for atlas generation.

All CTVs except those of the left-out dataset were mapped to atlas space using the existing deformation fields, and merged. The resulting contouring probability model was thresholded to 50% to obtain an auto-segmented CTV contour.

2.2. Validation

The technique was validated on a whole breast radiotherapy cohort containing datasets from 28 patients (14 right, 14 left), collected for a previous study [8]. Non-contrast computed tomography (CT) scans were used with whole breast clinical target volumes (CTV) delineated by eight independent observers. Observers followed a delineation protocol and were blind to other observer contours.

Since atlas accuracy correlates with similar anatomy of input patients, the cohort was divided into categories, grouping similar patients. Body mass index (BMI) was used as a surrogate for similar patient anatomy, with small/large patients classed as </> 29 BMI. Datasets were further divided according to laterality. An atlas was generated from the 6-9 datasets in each category using a leave-one-out approach. The atlas was registered using rigid and affine transformation followed by non-rigid registration, to the 'left out' dataset. The segmentation was mapped to this dataset from atlas space, and clipped such that it did not extend past the patient surface.

A gold standard, consensus contour was generated from the observer CTVs on the left-out dataset in each category using the Simultaneous and Truth Performance Level Estimation (STAPLE) algorithm [9]. To assess similarity, the gold standard contour was compared to the segmentation using the dice similarity coefficient (DSC) [10] and mean absolute surface distance (MASD) [11]. A comparison between the smallest and largest CTV with the segmentation was also made to determine the range in accuracy.

3. Results

Average atlases with corresponding whole breast CTV segmentations were successfully generated for each category as shown in figure 2 with corresponding probability maps.

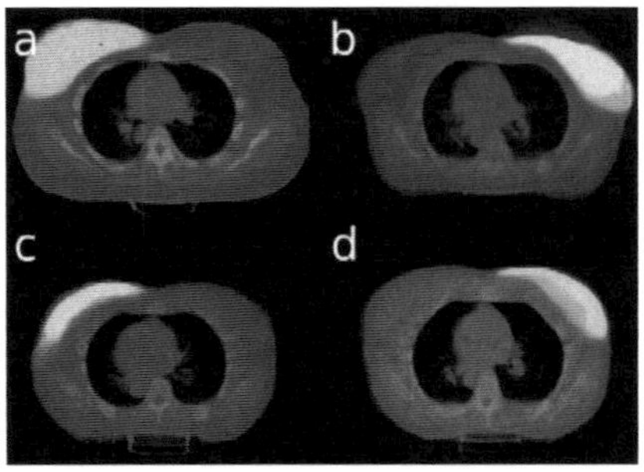

Figure 2. Probability maps of observer contours for a) large right, b) large left, c) small left, d) small right atlas'. Heat map indicates regions of highest probability.

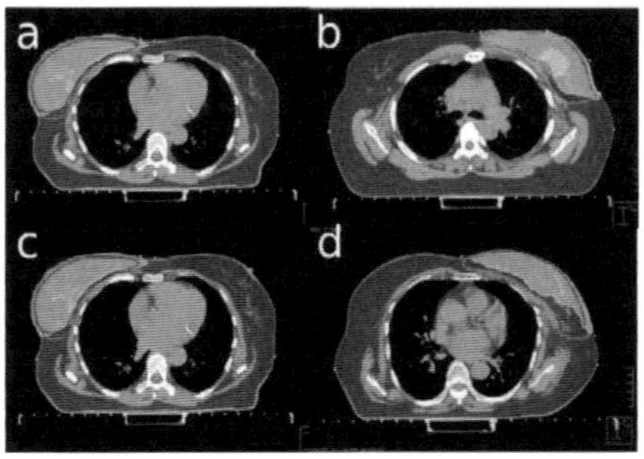

Figure 3. Segmentation (yellow) for a) large-right, b) large-left, c) small-left, d) small-right datasets. STAPLE (green), smallest (blue) and largest (red) CTVs are shown for comparison.

The whole breast CTV segmentation is visually compared to the consensus, smallest and largest contour in figure 3. The time required to segment, including registration of the atlas to the required patient and propagating contours, was on average 3 min and 43 seconds and didn't exceed 4 min and 49 seconds. The DSC and MASD values indicating the accuracy of the auto-segmentation in delineating the whole breast CTV and accounting for inter-observer variation is shown in table 2.

Table 2. Similarity metrics comparing whole breast segmentation with manual target volumes

	DSC				MASD (mm)			
	Large Left	Large Right	Small Left	Small Right	Large Left	Large Right	Small Right	Small Left
Consensus	0.81	0.85	0.71	0.79	7.99	3.47	6.17	5.16
Smallest CTV	0.77	0.86	0.71	0.79	9.28	3.64	4.81	5.16
Largest CTV	0.81	0.8	0.71	0.79	7.50	4.71	6.13	5.07

4. Discussion

Coverage probability maps overlaid on average atlas' showed high observer agreement for the majority of the whole breast segmentation. Observer agreement was lowest in the anterior-medial and posterior-lateral directions as is supported in the literature [12].

Segmented contours generally compared well with the gold standard for each category. Insufficiencies in segmentation were observed close to the chest wall for large and small BMI left breast patients. The authors hypothesise that this is due to larger contouring variations due to the close proximity of the heart. Manual review of this key area by a clinician is recommended. Improvements in segmentation in these instances could be achieved through optimisation of the probability map thresholding value.

5. Conclusion

This atlas based segmentation technique incorporating interobserver variation has been shown to be accurate (DSC >0.7, MASD <1cm) and efficient (time < 5 min) as is necessary for the clinical efficacy of adaptive radiotherapy. This successfully demonstrates an approach for presenting likely inter-observer variation determined from previous datasets on a current dataset. There is scope for further optimisation of segmentation correlation with gold standard contours as well as computation time. This method is a feasible solution to contouring inefficiencies for adaptive radiotherapy, with the potential for viable application to other anatomical sites.

6. Acknowledgments

The authors would like to acknowledge V.Batumalai and G.Delaney for the provision of data for this study. This research is funded by a Cancer Australia and National Breast Cancer Foundation (1033237), Liverpool Hospital Radiation Oncology trust fund, Australian Postgraduate Award and South Western Sydney Radiation Oncology Student Scholarship.

7. References

[1] Vial P *et al* 2008 *Med. Phys.* **35** 1267-77
[2] Reed V K *et al* 2009 *Int. J. Radiat. Oncol. Biol. Phys.* **73** 1493-500
[3] Njeh C F 2008 *J. Med. Phys.* **33** 136-40
[4] Dowling J A *et at* 2011 *Proc. Int. Workshop held in conjunction with MICCAI (Toronto, Canada, 22 September, 2011)* (Heidelberg: Springer) pp 10-21
[5] Dowling J A *et al* 2012 *Int. J. Radiat. Oncol. Biol. Phys.* **83** e5-11
[6] Ourselin S *et al* 2001 *Image Vision Comput.* **19** 25-31
[7] Vercauteren T *et al* 2009 *NeuroImage* **45** S61-72
[8] Batumalai V *et al* 2011 *Clin. Oncol.* **23** 108-13
[9] Warfield S K *et al* 2004 *IEEE T Med. Imaging* **23** 903-21
[10] Abreu *et al* 2006 *Proc. 12th Pacific Rim Int. Symp. on Dependable Computing (Riverside, California, 18-20 December, 2006)* (IEEE) pp 39-46
[11] Babalola K O *et al* 2009 *NeuroiImage* **47** 1435-47
[12] Pogson E M *et al* 2014 *J. Phys. Conf. Ser.* **489** 012057

Evaluation of dose from kV cone-beam computed tomography during radiotherapy: a comparison of methodologies

J Buckley[1], D Wilkinson[2], A Malaroda[1], P Metcalfe[1]
[1]Centre for Medical Radiation Physics, University of Wollongong, New South Wales 2522, Australia
[2]Illawarra Cancer Care Centre, Wollongong Hospital, New South Wales 2500, Australia

E-mail: jarryd@uow.edu.au

Abstract. Three alternative methodologies to the Computed-Tomography Dose Index for the evaluation of Cone-Beam Computed Tomography dose are compared, the Cone-Beam Dose Index, IAEA Human Health Report No. 5 recommended methodology and the AAPM Task Group 111 recommended methodology. The protocols were evaluated for Pelvis and Thorax scan modes on Varian® On-Board Imager and Truebeam kV XI imaging systems. The weighted planar average dose was highest for the AAPM methodology across all scans, with the CBDI being the second highest overall. A 17.96% and 1.14% decrease from the TG-111 protocol to the IAEA and CBDI protocols for the Pelvis mode and 18.15% and 13.10% decrease for the Thorax mode were observed for the XI system. For the OBI system, the variation was 16.46% and 7.14% for Pelvis mode and 15.93% to the CBDI protocol in Thorax mode respectively.

1. Introduction

The introduction of kV CBCT for position verification and adaptive therapy has enabled steeper dose gradients and tighter treatment margins for improved treatment outcomes [1]. However, regular CBCT imaging during a course of radiotherapy delivers additional concomitant dose which may lead to increased risk of secondary malignancies [2-4].

The current paradigm for evaluating CT dose is the Computed Tomography Dose Index [5-7]. In its most common form, the integral dose from a single rotation of the X-Ray tube is measured with a 100mm pencil ionisation chamber and, when divided by the slice width nT gives an estimate of the central dose for a scan with multiple rotations of the X-Ray tube. Since the introduction of CTDI, CT scanning has developed to include helical CT, multi-slice CT and now wide beam Cone-Beam CT where images can be acquired in a single rotation. Hence, the relevance of the CTDI as a dose indicator for wide beam scanning has come under question due to underestimation of scatter dose lying outside the 100mm chamber length, CTDI phantoms being of insufficient length to achieve scatter equilibrium and non-uniformity along the beam [8-10].

In an attempt to better quantify the dose from wide-beam scanning, alternative methodologies have been developed which attempt to counter the limitations of the CTDI methodology. This work presents a comparison of three alternative methods, the Cone-Beam Dose Index, IAEA Report 5 Recommended Methodology and the AAPM Task Group 111 protocol [11-13].

2. Materials and Methods

The methodologies were compared on Varian© On Board Imager (OBI) and XI kV imaging systems. Both systems consist of a kV X-Ray tube mounted on a LINAC gantry 90 degrees to the treatment beam, with a flat panel detector 180 degrees to the X-Ray tube from which 3D images can be reconstructed.

2.1 Cone Beam Dose Index (CBDI)

The CBDI was evaluated with a 100mm ionisation chamber for a single rotation of the X-Ray tube. The dose was measured in-air at the centre of a Polymethyl-Methacrylate cylindrical phantom, 32 cm in diameter and 16 cm in length.

Unlike the CTDI, the integral dose D(z) was not divided by the slice thickness, which for CBCT is wider than the chamber itself, but by the chambers 100mm length. That is, the 100mm chamber records a point-dose measurement as shown in equation (1).

$$CBCT = \frac{1}{100mm}\int_{-50mm}^{+50mm} D(z) \qquad (1)$$

The weighted average dose across the phantom plane was determined by weighting the central dose and the average peripheral dose shown in equation (2), as per CTDIw in the CTDI. The CTDIw was evaluated for Pelvis and Thorax modes on both the OBI and XI systems.

$$CTDI_w = \frac{1}{3}CTDI_c + \frac{2}{3}CTDI_p \qquad (2)$$

2.2 IAEA Report 5 Recommended Methodology

The IAEA protocol was carried out following the recommendations of the International Electrotechinical Commission -60601-2-44 document. The weighted average dose was determined for a reference beam width of nT = 20mm for the Pelvis and Thorax modes on the OBI and XI systems.

The weighted average dose was determined free-in-air following the guidelines of the IAEA for the reference and Pelvis/Thorax protocol beam widths. A single measurement was taken for the reference width with the chamber in the central position, while for the wide protocol beam width the dose was measured in three sequential steps of the chamber through the beam and summed as shown in equation (3).

$$CTDI_{free\ in-air} = \frac{L_c}{nT}\sum_{3}^{i=1} D_i \qquad (3)$$

The CTDIIAEA was then evaluated by multiplying the weighted average dose from the reference width scan by the ratio of the CTDIfree in-air values of the reference and protocol width scans as shown in equation (4).

$$CTDI_{free\ in-air} = \frac{L_c}{nT}\sum_{3}^{i=1} D_i \qquad (4)$$

2.3 AAPM Task Group 111 Methodology

A custom made PMMA phantom was constructed in house following the TG-111 recommendations for phantom length to achieve scatter equilibrium in the centre of the phantom. The phantom was cylindrical of diameter 32cm with 5 plug housings for weighted average measurements. The phantom was 45cm in length.

Measurements were taken in each position with a 2571 0.6cc Farmer ionisation chamber. The plug containing the chamber was custom made to conform to the chamber geometry to eliminate air gaps. The weighted average dose was determined for Pelvis and Thorax protocols on the OBI and XI systems using the chamber readings (converted to dose in-air) and equation (2). The charge readings from the ionisation chamber were converted to dose to air following the AAPM TG-61 guidelines [14].

3. Results

Weighted average dose values for OBI and XI are shown in figure 1 (a) and (b). Measurements for each position within the phantoms are shown in figure 2(c)-(f).

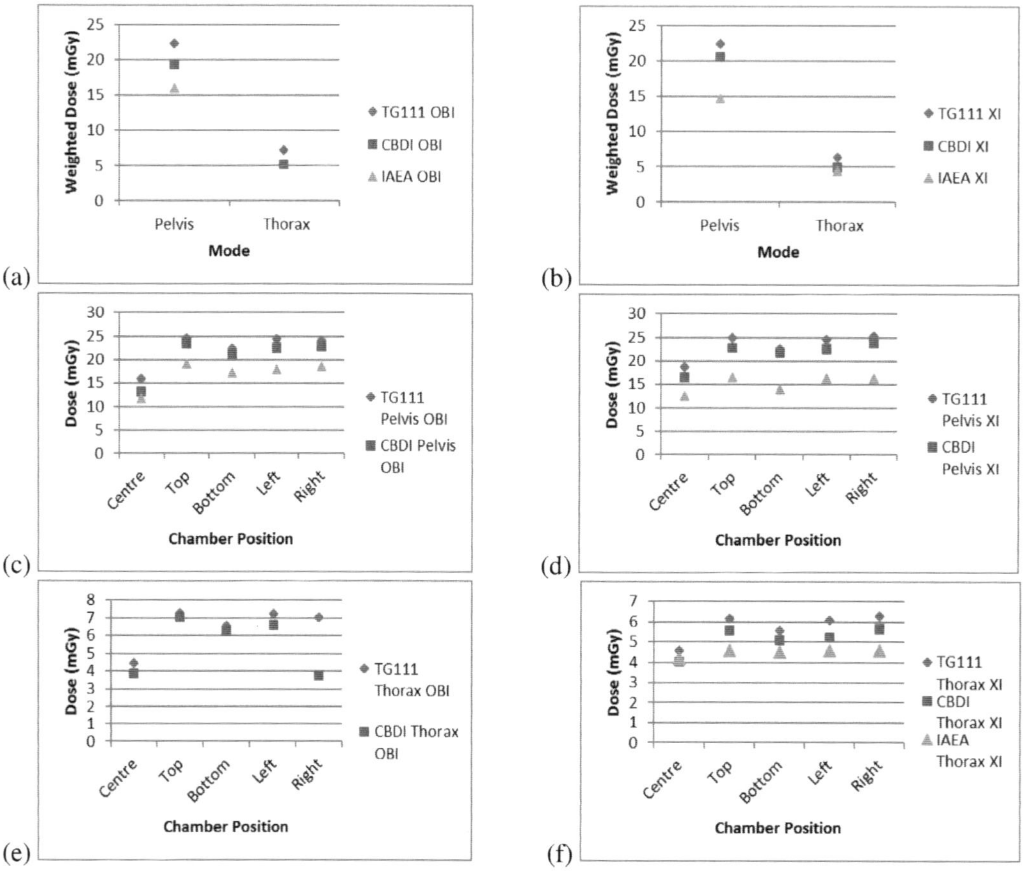

Figure 1. Positional dose values for the Pelvis and Thorax modes on the OBI (c,e) and XI (d,f) imaging systems. The measurements were taken following CBCT, IAEA and TG-111 recommendations. The weighted averages for the three methodologies are given for OBI (a) and XI (b).

Figure 1. Positional dose values for the Pelvis and Thorax modes on the OBI (c,e) and XI (d,f) imaging systems. The measurements were taken following CBCT, IAEA and TG-111 recommendations. The weighted averages for the three methodologies are given for OBI (a) and XI (b).

Across all imaging systems, the TG-111 protocol gave the highest dose due to the increased scatter from the longer phantom. The CBCT methodology gives slightly lower dose across all protocols, with the exception of the right position in the OBI pelvis scan which is considerably less, likely due to a delay between the X-Ray tube starting and the gantry beginning its rotation, with the tube directly above the chamber. The central measurement for the TG-111 is considerably lower than the other two methods for the Pelvis XI scan, perhaps due to the smaller chamber volume not collecting as much scatter dose compared to the longer chamber, or misalignment of the chamber, which would have a more pronounced effect on the smaller volume chamber. Measurements in the bottom position were all slightly less than the other peripheral positions due to couch attenuation.

Weighted dose values for each protocol are shown in figure 2 (a) for OBI and (b) for XI. For the XI system there was a 17.96% and 1.14% variation from the TG-111 protocol to the IAEA and CBDI protocols for the Pelvis mode, while the Thorax mode had 18.15% and 13.10% variations. The OBI variation was 16.46% and 7.14% for Pelvis mode and 15.93% to the CBDI protocol in Thorax mode.

4. Conclusion

Three alternative methodologies were compared on two CBCT imaging systems. The TG-111 protocol gave the highest dose value across all scan modes on both the OBI and XI imaging systems due to the increased scatter from the longer custom phantom. The CBDI method gave the second highest dose, while the IAEA method gave the lowest dose. Across all scans, the central dose was lower than the peripheral values due to the primary beam contributing less, and the scatter dose contributing more relative to the peripheral positions.

The TG-111 protocol, with its higher dose due to increased scatter length gives a more realistic indication of dose from wide-beam CBCT scanning. Future work is required to investigate how the TG-111 values correspond to patient dose, including the influence of patient thickness.

5. References

[1] Boda-Heggermann J *et al* 2011 *J. Strah. Onkol.* **187** 284
[2] Kim D *et al* 2013 *Health. Phys.* **104** 999
[3] Kim S *et al* 2010 *A. J. Roentgenol.* **194** 186
[4] Greer P B *et al* 2007 *Med. Phys.* **34** 4389-98
[5] Hill B *et al* 2005 *Med. Phys.* **32** 1589-97
[6] Hill B *et al* 2008 *Physica Medica* **24** 149-58
[7] Shope T *et al* 1981 *Med. Phys.* **8** 488
[8] Boone J 2007 *Med. Phys.* **34** 1364
[9] Dixon R 2003 *Med. Phys.* **30** 1272
[10] Zhou H and Boone J 2008 *Med. Phys.* **35** 2424
[11] Amer A *et al* 2007 *Br. J. Radiol.* **80** 476
[12] IAEA 2011 Human Health Reports Series No. STI/PUB/1528 (Vienna: International Atomic Energy Agency)
[13] AAPM 2010 Task Group Report 111 (MD: American Association of Physicists in Medicine)
[14] Ma C *et al* 2001 *Med. Phys.* **28** 868

Out of field dose during Gamma Knife treatment: a paediatric case study

V Moutrie[1], M Grace[1], M A Izard[1], J W Fuller[2]
[1]Genesis Cancer Care, Suite 1 Level B2, 3 Technology Place, Macquarie University, NSW 2109 Australia
[2]Macquarie Neurosurgery, Suite 201, Level 2, 2 Technology Place, Macquarie University NSW 2109 Australia

E-mail: vaughan.moutrie@genesiscare.com.au

Abstract. An 11-year-old girl with an arteriovenous malformation (AVM) was referred for Gamma Knife treatment. As this would be the first paediatric treatment in Australia, additional investigations were undertaken into out of field dose to assure the best possible long term outcome for the patient. A phantom was constructed from water equivalent materials to simulate the patient. A target volume was defined to emulate the size and location of the AVM visible in diagnostic images. An ionisation chamber and EBT3 Gafchromic film were used to record absorbed dose at strategic points both on the surface and at depth within the phantom. On the day of treatment, EBT3 Gafchromic film was used to conduct in vivo dosimetry. The pre-treatment phantom measurements matched the planning system for the cranial section (the only modelled section) and no measurable dose above background was detected in the extracranial sites. In vivo measurements of the lenses returned doses of up to 2 cGy for imaging and 8 cGy for treatment which was also consistent with the planned dose. Dose to the thyroid, chest and abdomen was not measurable above background.

1. Introduction

It is important when delivering radiation therapy that dose outside the intended target volume is kept as low as possible to reduce unwanted side effects [1]. This is especially true for stereotactic techniques, where large doses are delivered in hypofractionated or single fraction settings. Out-of-field dose minimisation is particularly important for paediatric patients for several reasons: increased consideration towards radiation induced malignancies due to long life expectancy, relatively limited information on paediatric radiobiology and, in the case of intra-cranial treatments, the generally shorter distance between the target area and radiosensitive organs.

An 11-year-old girl with an arteriovenous mal-formation (AVM) was referred for Gamma Knife treatment. Because of the considerations above, it was decided to conduct a detailed assessment into the potential delivered dose to the patient prior to proceeding with the treatment, and monitor the dose received throughout the course of the procedure.

A literature review found several previous studies [2-5] investigating peripheral doses in Gamma Knife treatment, some including paediatric cases [2, 4], however all of these studies were performed using previous generation helmet-based delivery systems that differ from the Perfexion machine at our facility and are likely to have different out-of-field dose levels. These studies used a mix of in vivo patient dosimetry and phantom measurements using ion chambers and thermoluminescent detectors.

Extracranial doses across the studies were in the order of 0-0.2 Gy. It was noted [4] that measured organ doses varied significantly based on the exact treatment conditions.

2. Methods

2.1. Preliminary study

A phantom was constructed from water equivalent materials to simulate the patient (figure 1) using self-reported physical dimensions from the patient. Calibrated ionisation chambers and EBT3 Gafchromic film were used to record absorbed dose at strategic points both on the surface and at depth within the phantom. As only surface measurements can be made in vivo, this allowed validation of measurement points on the surface as good representations of corresponding organs at depth in the same location.

Figure 1. Configuration of the test phantom.

A spherical dosimetry phantom was used to simulate the head of the patient and computed tomography (CT) image set used to plan the mock treatment. A PTW 31010 Semiflex ion chamber was inserted to the centre of the sphere. Pieces of EBT3 film were placed in the position of the eyes. A sheet of jelly bolus was rolled into a cylinder to simulate the neck and a piece of film taped to the surface to measure thyroid dose. The remainder of the body was simulated with stacks of Solid Water sheets. A PTW 31013 Farmer ion chamber was placed at a depth of 1.5 cm in the centre of superior block to measure sternum dose. EBT3 film was placed in positions to measure breast and ovary dose both at the surface and at a relevant depth (0.5 cm and 8 cm respectively). The position of the breast/sternum was approximately 30 cm from the centre of the head and the ovary position a further 30 cm inferiorly.

A target volume was defined to emulate the size and location of the AVM visible in diagnostic images (figure 2), and a dose of 100 Gy to the 50% isodose, five-fold higher than the regular prescription, was planned to ensure adequate signal to the dosimeters. The predicted dose to the centre of the sphere was 0.6 Gy, the right eye was 1.4 Gy and the left eye 1.6 Gy. As very low readings were anticipated, background readings were measured for each ion chamber for a period of 9999 s prior to delivery and used to correct the measured reading. Film was processed according to the standard in vivo dosimetry procedure used at the facility [6].

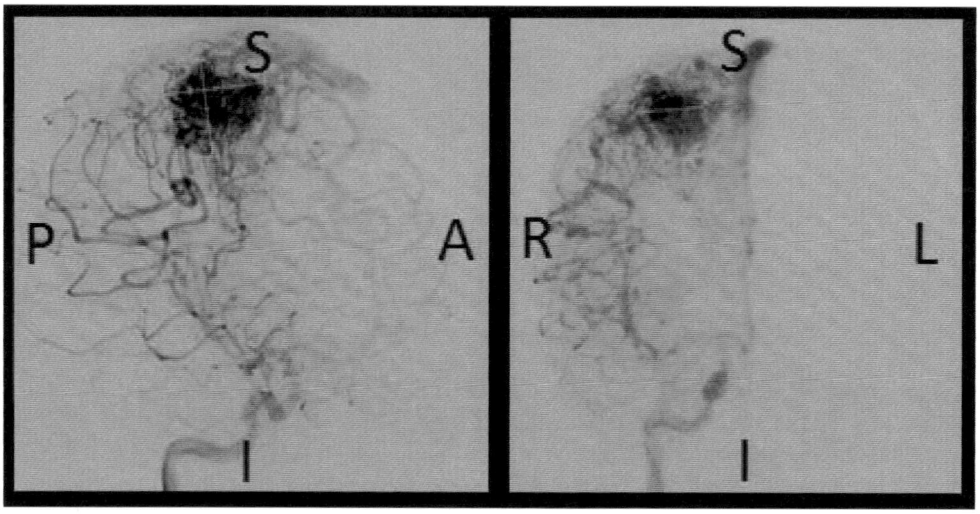

Figure 2. Location of the AVM on early diagnostic scans.

2.2. In vivo dosimetry

In vivo dosimetry was conducted on the day of treatment, separately monitoring imaging (digital subtraction angiography (DSA), CT) and treatment dose to the eyes, chest, abdomen, and for treatment only, thyroid. The treatment prescription was 20 Gy to the 50% isodose.

3. Results

Table 1. Results for the preliminary study, peak dose 200 Gy.

Dosimeter	Site	Location	Measured dose (Gy)	Planned dose (Gy)
Semiflex	Centre of head	Centre of sphere	0.5	0.6
Farmer	Sternum	Midline, 1.5 cm deep	b	n
Film A	Right eye	4 cm lat, surface	1.0	1.4
Film B	Left eye	4 cm lat, surface	1.7	1.6
Film C	Thyroid	Midline, surface	0.04	n
Film D	Right breast	10 cm lat, 0.5 cm deep	0.02	n
Film E	Left breast	10 cm lat, 0.5 cm deep	0.02	n
Film F	Right breast	10 cm lat, surface	0.03	n
Film G	Left breast	10 cm lat, surface	0.03	n
Film H	Ovary	Midline, 8 cm deep	b	n
Film I	Ovary	Midline, surface	b	n

b: Measurement was indistinguishable above background, <0.01.
n: Planned dose not available as location is outside the region bound by the stereotactic frame.

Table 2. Results for the in vivo dosimetry, peak dose 40 Gy.

Site	Imaging dose (Gy)	Treatment dose (Gy)	Planned treatment dose (Gy)
Left eye	0.01	0.06	0.1
Right eye	0.02	0.08	0.1
Chest	b	b	n
Abdomen	b	b	n
Thyroid	-	b	n

b: Measurement was indistinguishable above background, <0.01.
n: Planned dose not available as location is outside the region bound by the stereotactic frame.

4. Discussion

All extracranial measurement points showed exceedingly small dose, even under experimental conditions where an escalated prescription dose was used. Reasonable agreement between planned and measured dose was found for points within the dose calculation volume.

In vivo measurements only detected dose to the eyes, with no distinguishable dose above background for the thyroid, chest or abdomen. Dose received from the CT and DSA was just discernible at 1 cGy and 2 cGy for the left and right eyes respectively. The higher dose to the right eye is rationalised by the placement of the x-ray source on this side for lateral imaging during DSA. The right eye also received slightly higher dose during treatment due to being nearer the target volume seen in figure 2. The point dose tool in the treatment planning system reports dose to the nearest 0.1 Gy, and in the vicinity of the eyes the displayed dose varied between 0.0 Gy and 0.1 Gy. This predicted dose is therefore a match to the measured doses of 0.06 Gy and 0.08 Gy for the left and right eyes respectively to within the associated degree of uncertainty.

5. Conclusion

The rapid drop in dose outside the field was verified and measured doses to organs at risk were very small. The study demonstrates the capability of the Gamma Knife to confine high doses to the intended volume, and the accuracy of the planning system in predicting out-of-field dose.

6. References

[1] Greer P B *et al* 2007 *Med. Phys.* **34** 4389-98
[2] Berk H W 1993 *Stereotact. Funct. Neurosurg.* **61** 164-72
[3] Ioffe V 2002 *Surg. Neuro.l* **58** 32-41
[4] Miljanic S 2013 *Radiat. Meas.* **55** 38-42
[5] Hasanzadeh H 2006 *Phys. Med. Biol.* **51** 4375-83
[6] Moylan R 2013 *Australas. Phys. Eng. Sci. Med.* **36** 331-7

Linearization of EBT3 film dose response and virtual film dosimetry for SBRT quality assurance

M Cai[1], B Archibald-Heeren[1,2], Y Wang[2], P Metcalfe[1]
[1]Center of Medical Radiation Physics, University of Wollongong, Northfields Ave, Wollongong NSW 2522, Australia
[2]Physics department, Radiation Oncology Institute, 185 Fox Valley Rd, Wahroonga NSW 2076, Australia

E-mail: mc267@uowmail.edu.au

Abstract. EBT3 film offers high spatial resolution and low energy dependence, making it a suitable choice for quality assurance where high dose gradients are present, such as the case for SBRT. This work presents a simple method to adjust scanner settings so that dose response becomes linear. This linearity eliminates the need to obtain a calibration curve and associated uncertainties in curve fitting. Relative dosimetry can be performed after dose normalization to a reference point. Linearity is also a more robust condition than calibration curve with respect to scanner warm-up conditions, resulting in reduced uncertainty in dose measurement. An in-house developed program reads the film scan and a 2D dose map then constructs both to virtual films using grayscale values. Film intensity value was normalized to dose at reference point. Relative dosimetry was performed by comparing the two resulting images. Patient specific quality assurance was conducted for two SBRT cases. In both plans more than 95% gamma function points passed the gamma criteria of 2%/3mm.

1. Introduction

Stereotactic body radiation therapy (SBRT) uses high dose radiation to target and destroy a well-defined tumour. In such procedures high dose gradients are often present to spare the surrounding healthy tissues from radiation damage. Patient specific quality assurance (QA) is an important step in making sure the planned dose is deliverable and the desired treatment outcome can be achieved [1-3]. Radiochromic film dosimeter such as EBT3 film offers high spatial resolution, near tissue equivalence and low energy dependence, making it a suitable choice for SBRT QA. A typical film dosimetry system requires establishing a calibration curve which converts change in optical density (OD) to absolute dose [4]. This dose response is often non-linear and the calibration curve is sensitive to changes in scanning conditions, thus increasing errors in dose measurement and limiting its use to relative dosimetry. The often lengthy calibration and scanning protocol also discourages its use in routine QA [5]. In this work, a simple method is presented where the scanner settings are adjusted such that scanned film intensity value becomes linear with respect to dose. Subsequently, intensity values can be used directly for relative dosimetry without calculating dose first. Delivered dose at a reference point, represented by its intensity value, can be assumed to equal planned dose - an assumption that can be easily verified with an ion chamber. Delivered dose at other points on the film can then be normalized and compared with planned dose directly. This will greatly simplify the film dosimetry procedure. Another advantage is that as linearity method does not rely on establishing a calibration curve, uncertainties in curve fitting parameters are eliminated. Linearity can also be a more invariable condition than a calibration curve with respect to variations in scanning conditions, resulting in reduced uncertainties in dose measurement.

2. Materials and methods

2.1. Establishing linearity

One 8'' x 10'' sheet of EBT3 film was cut into 16 strips along the vertical axis. Each strip measures 20.3cm x 1.59cm. The top margin of each film strip was marked with a number from 0 to 1500 in 100 increments indicating the dose (in cGy) to be delivered. The film strips were irradiated in a 6MV photon beam from a Varian Clinac linear accelerator, the machine output of which has been verified using IAEA protocol [6]. During irradiation, each film was placed at a source-to-surface distance (SSD) of 100cm inside a solid water phantom. A 10cm slab was placed on top of the film and at least 10cm slab was placed underneath to provide sufficient backscatter. A field size of 10cm x 10cm was used and each film strip was placed at the center of the field. Film was handled according to TG-55 protocol [7].

An Epson Expression 10000XL flatbed scanner and EpsonScan software were used to scan the films. The glass plates of the scanner were sanitized prior to scanning. Irradiated film strips were placed side by side in increasing order of radiation exposure and taped together at top and bottom to form one sheet, as shown in figure 1. It was then placed inside the scanner with the long axis along the scanning direction to reduce scanner light scattering effect [8]. In the EpsonScan software, transmission mode was selected and all filters and image enhancement options were turned off. The film was scanned in both 48-bit color and 16-bit grayscale mode and saved in TIFF format. All of the red, green, blue channels and the grayscale values were extracted and analyzed separately. A rectangle region of interest (ROI) of 10 pixel x 100 pixel was selected from the center of each strip and the average grayscale value (GSV) was used as the intensity reading for that given dose. An unexposed film was scanned and the ROIs in the same positions were selected to correct for non-uniformity.

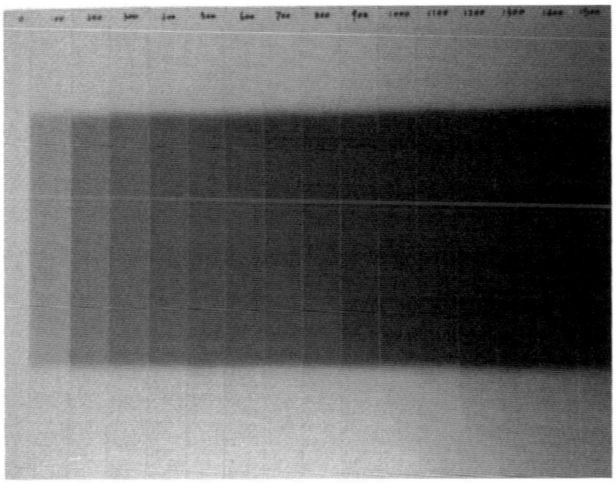

Figure 1. Calibration film strips prepared for scanning.

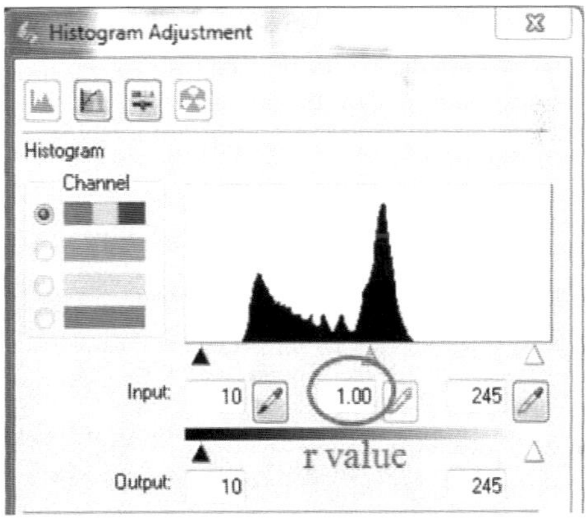

Figure 2. Histogram adjustment tab of the EpsonScan software.

Dose response was modelled with a straight line through origin:

$$\text{Dose} = a \cdot netGSV = a \cdot (GSV_{\text{exp}} - GSV_{\text{unexp}})$$

Linearity was evaluated by the R-square value of the fit. In order to linearize the dose response, EpsonScan settings were investigated. We aim to find an adjustment parameter that can be quantified and can alter the shape of the dose response. Preliminary studies had thus narrowed down the choice of adjustment parameter to the gamma correction value (referred to as the r value in this work) in the histogram adjustment tab of the EpsonScan software, as shown in figure 2. The r value is the equivalent of the inverse of gamma value in image processing. The following r values were tested: 0.5, 1 (default), 2, 3, 4, 5 and 6. A scan was performed after each change of setting. An in-house developed code written in MATLAB was used to select the ROI in each dose region, calculate the netGSV, plot the dose response and compute the R-square value.

In order to test the robustness of linearity with respect to scanner warm-up time, films were scanned immediately after start-up then after 30 minutes and 60 minutes of warm-up time.

2.2. Relative dosimetry for SBRT patient QA

Two SBRT treatment plans of the lung and head and neck region were generated using RayStation treatment planning system. For each plan, one slice was selected from the planning 3D dose distribution and exported as a 2D dose map. The same slice was delivered to an EBT3 film from the same batch as the calibration film in a cylindrical IMRT QA phantom. Irradiated film was scanned using scanner settings established at linearity condition. An in-house developed software loads the scanned film and 2D dose map as two matrices of intensity values. Both matrices were reconstructed to virtual films for display and visual inspection. The software autoregisters the two images based on film dimensions. One point at the edge of the film and one reference point (e.g. ISO center) were selected to normalize the film intensity to dose. The resulting film image can then be compared directly with dose image for relative dosimetry. Criteria for gamma-value analysis was set at 2%/3mm.

3. Results

Auto-histogram option in EpsonScan was selected as it utilizes all available grayscale channels and does not alter linearity. A linear dose response could be established by setting r value to 4. Lower values were insufficient to correct non-linearity while higher values result in heavy compression of intensity output. Figure 3 shows the dose response from each of the color channels as well as the grayscale image. Both grayscale and green channel provide good linearity in the dose range considered. As grayscale image requires only one third of storage space compared to a color image and no extraction of color channel is required, it is the preferred choice for clinical dosimetry. For this reason, all subsequent analysis is based on grayscale image only.

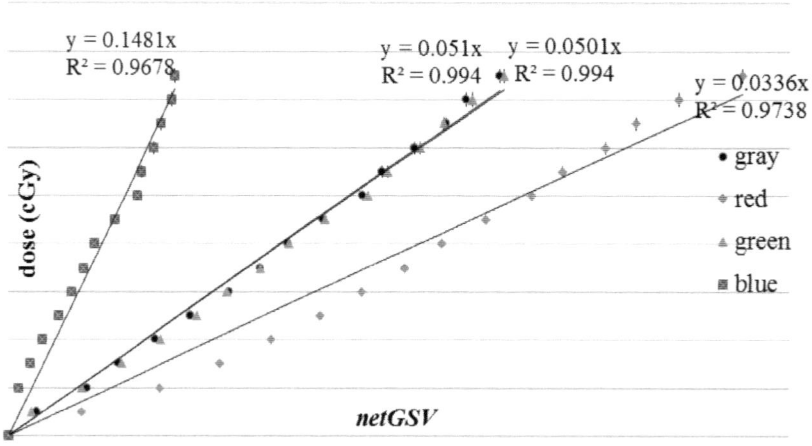

Figure 3. Dose response using grayscale and each colour channel.

When scanner warm-up time was varied from zero through to 60 minutes, the linear coefficient changed slightly due to variations in intensity values but linearity remained the same. In clinical applications, two main factors affecting dosimetry efficiency are the wait time post-irradiation for the film to develop and the recalibration for each film batch. Future work will test the robustness of linearity with regards to film development time and different film batches.

Linearity method reduces uncertainties in dose measurement in two ways. Firstly, the only parameter of concern is linearity and the linear coefficient is irrelevant. Thus the uncertainties in curve fitting parameters are eliminated. Moreover, with traditional OD method, once a film batch is calibrated, subsequent processing and scanning of the film requires following a protocol aimed at reproducing the condition under which calibration occurred. This is because any change in such condition will result in a shift of the calibration curve, and subsequent dose measurement. With the linearity method, as linearity is insensitive to variations in scanning conditions such as scanner warm up time, any change in intensity values will be self-corrected when intensity values are normalized for relative dosimetry.

Linearity method was applied for SBRT patient QA. Exported dose maps have 1mm spatial resolution. When used for dosimetry analysis they provide detailed dose information which is particularly important in testing high dose gradient area. In both tested plans, more than 95% of gamma function points passed gamma-value criteria of 2%/3mm. The film dosimetry process is straightforward and efficient.

4. Conclusions

It was demonstrated that a robust linearity of dose response for EBT3 film dosimeter up to 15Gy can be established by adjusting scanner settings. As a result, film dosimetry can be conducted without finding a calibration curve beforehand and without regard to scanner warm-up time. An in-house developed software was used to conduct relative dosimetry and proved to provide a convenient way of performing routine QA for SBRT cases with high resolution and accuracy.

5. References

[1] Vial P *et al* 2008 *Med. Phys.* **35** 1267-77
[2] Greer P B *et al* 2007 *Med. Phys.* **34** 4389-98
[3] Hill R *et al* 2014 *Med. Phys.* **36** R183-231
[4] Devic S 2005 *Med. Phys.* **32** 2245-53
[5] Hill R *et al* 2010 *Med. Phys.* **37** 4355-63
[6] IAEA Technical Reports Series No. 398 2000 (Vienna: International Atomic Energy Agency)
[7] Niroomand-Rad A 1998 55 *Med. Phys.* **25** 2093-115
[8] Martisíková M 2008 *Phys. Med. Biol.* **53** 7013-27

Applications of a superconducting solenoidal separator in the experimental investigation of nuclear reactions

D J Hinde, I P Carter, M Dasgupta, E C Simpson, K J Cook, Sunil Kalkal, D H Luong, E Williams
Department of Nuclear Physics, Research School of Physics and Engineering, Australian National University, ACT 2601, Australia

E-mail: david.hinde@anu.edu.au

Abstract. This paper describes applications of a novel superconducting solenoidal separator, with magnetic fields up to 8 Tesla, for studies of nuclear reactions using the Heavy Ion Accelerator Facility at the Australian National University.

1. Introduction
Heavy ion beams are used at the ANU for nuclear reactions studies, with beam species typically ranging from from Li to Ni. These carry substantial linear momentum, meaning that many reaction products are focused around the beam direction. To study the reaction process, or make use of these reaction products, it is often necessary to separate them from the intense background of elastically scattered beam particles, which can be a million times more intense. To make efficient use of the beam, it is desirable to accept reaction products into a large solid angle.

2. Superconducting solenoid
These two goals can be achieved by placing a superconducting solenoid at zero degrees, with the unscattered beam being stopped before entering, and scattered beam particles and reaction products of interest being separated by their different magnetic rigidity, as they pass through the high magnetic

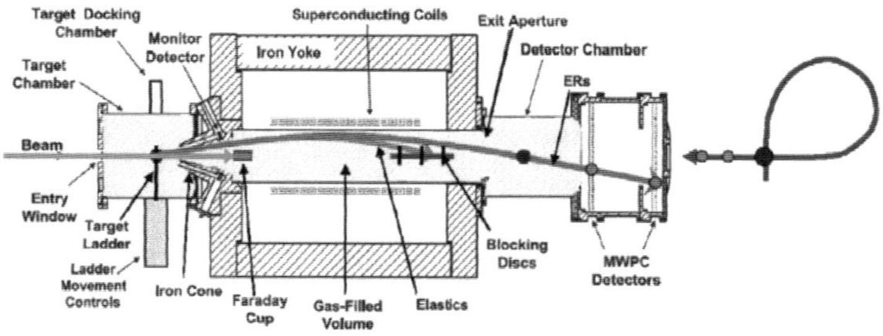

Figure 1. Schematic diagram of SOLITAIRE configured for fusion measurements. A fusion trajectory (blue) is superimposed showing radial separation from the axis (left), and in the perpendicular plane (right). The region filled with 0.7 mbar He gas is indicated in green.

Figure 2. Setting up SOLITAIRE for fusion measurements at the ANU.

fields in the solenoid. This gives helical paths with a focus on axis after the solenoid, with different focal lengths for different magnetic rigidities. By selection of the magnetic field, particles of interest can be transmitted, whilst beam particles are stopped inside the bore of the solenoid, as shown in figure 1. The device at ANU is called SOLITAIRE [1], which is shown in figures 2.

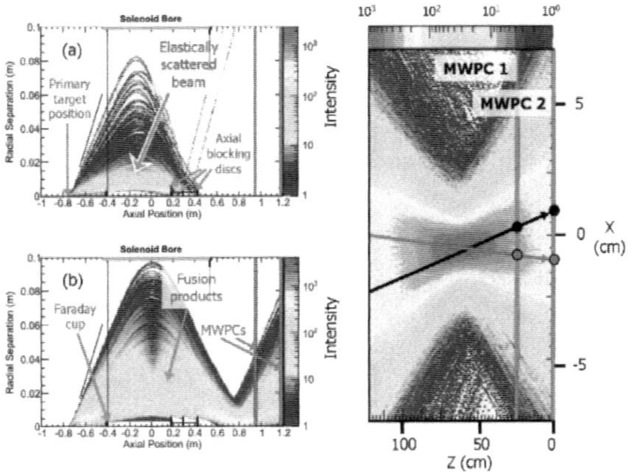

Figure 3. The left panels show simulated trajectories of elastic scattering and fusion products through SOLITAIRE. Beam particles are stopped, whilst fusion products come to a focus and are detected in the two MWPC detectors (see text). The right panel shows experimental trajectories determined by extrapolating the positions measured in the two MWPC detectors (magenta bars).

3. Application to nuclear fusion

Nuclear fusion products carry all the beam momentum. They are deflected from the beam direction by the generally much smaller recoil momentum imparted by evaporated light particles, emitted as they cool to form evaporation residues (ERs). For thick targets, angular scattering in the target can also spread the angles significantly. Nevertheless, fusion products are generally found inside a cone with 10 degree half-angle, centred around zero degrees [2]. Filling the solenoid with low pressure gas effectively collapses the broad fusion product charge state distribution to that of the mean charge state, giving a compact focal point. Figure 3 shows Monte Carlo simulations of particle trajectories, including scattering and charge exchange in the gas, for beam particles (top left) which are stopped on the axial discs, and fusion products (bottom left) which come to a focus before the two Multi Wire Proportional Counters (MWPCs). These detectors are position sensitive, with 1 mm resolution, thus together they can be used to determine the trajectories. Experimental fusion product trajectories in figure 3 (right) show that the focal length is well-determined, allowing the simulation to be matched to the experiment (by adjusting the fusion product mean charge state in the model). To obtain the transmission efficiency, knowledge of the angular distribution from the target is also required, as well as the trajectories through the solenoid. Figure 4 shows the XY position spectrum of fusion products, identified by their energy loss and time information from the MWPCs. From this, information on the angular distribution from the target can be determined, and thus the SOLITAIRE transmission efficiency to the detectors [1]. This allows high precision measurements of fusion cross-sections, essential to give a unique insight into the large effects of quantum coherence [2,3] (and also decoherence and energy dissipation [4]) that controls quantum tunnelling in fusion.

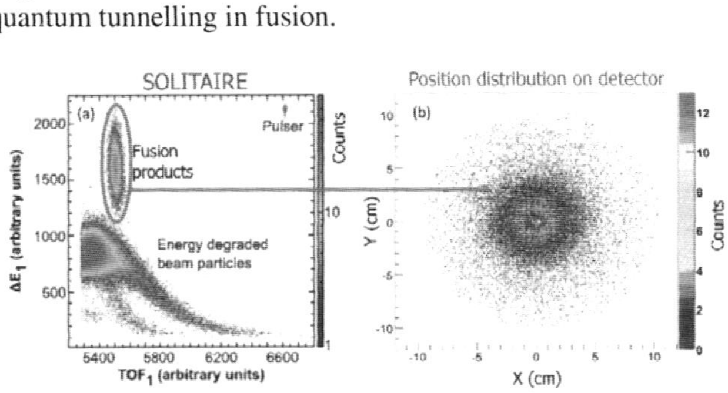

Figure 4. Identification of fusion products from the $^{34}S+^{89}Y$ reaction by energy loss (ΔE) and time-of-flight (TOF) from the target to MWPC detector #1. The position spectrum is shown on the right. The width of the image is related to the mean charge state and the shape of the angular distribution from the target before the solenoid.

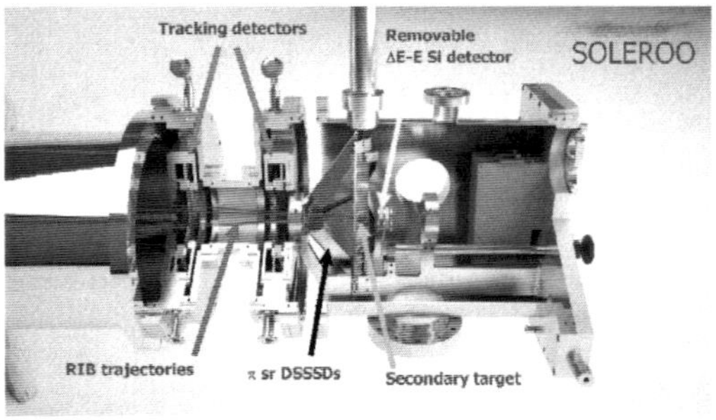

Figure 5. Computer rendering of the secondary target chamber of the SOLEROO radioactive beam facility. Tracking detectors give energy loss, position and timing information about the radioactive ion beam (RIB) particles (red) before they hit the secondary target. This identifies the type of particle, its energy, and impact point and angle of incidence onto the secondary target. A removable Si ΔE-E detector telescope allows checking of beam characteristics (see text).

4. Application to producing radioactive ion beams (RIB)

Radioactive nuclei far from stability are expected to have different properties from stable nuclei. In particular, nuclei with extreme neutron-to-proton ratios show neutron or proton halos (neutron or proton wave-functions extending far from the core) because of their low binding energy. The interactions of halo nuclei may show very different behaviour from well-bound nuclei [5]. It has been suggested that irreversible interactions of the halo nucleon(s) with those in the target nucleus [6] may be important, even at large separations, in contrast with normal expectations for well-bound nuclei.

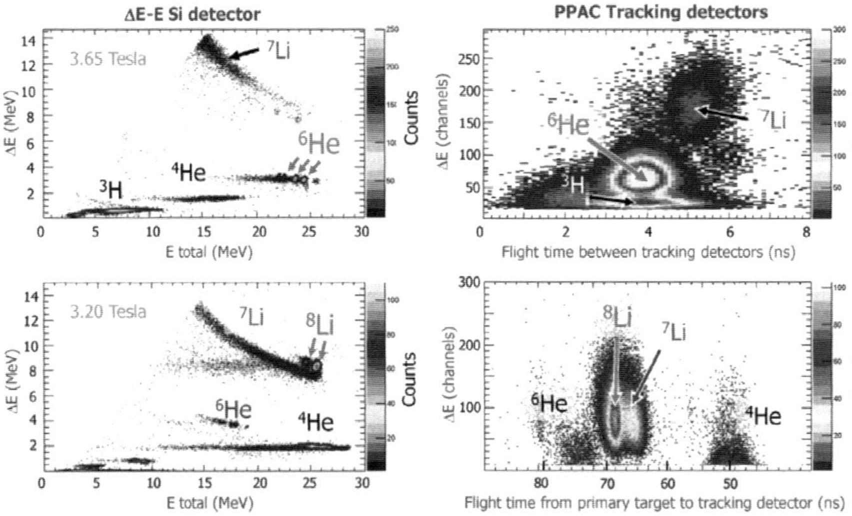

Figure 6. The left panels show spectra from the ΔE-E detector at the secondary target position, for two magnetic fields. The desired exotic beams (with two or more discrete energies) are indicated in red. The right panels show the spectra obtained (at higher beam intensity) derived from the ΔE and timing signals from the thin gas PPAC tracking detectors. The desired products can be cleanly tagged by the information from the tracking detectors.

The development of SOLITAIRE for RIB production (SOLEROO) has been described in [7, 8]. The radioactive nuclei are produced by nuclear reactions upstream of the solenoid, and the purified beam of interest is focused onto a secondary target downstream (shown in figure 5). A key quantity in precision measurements is beam purity and energy definition. To check the beam impurities at low intensity, the secondary target can be removed, and a Si ΔE-E telescope inserted. To identify impurities, two thin gas proportional detectors (PPACs) are located after the solenoid, but before the secondary target. These are capable of operating at count-rates over 10^6/sec. Triggered by a signal in one of the detectors viewing the secondary target, the PPACs record the time, energy loss and position of the ion generating the trigger. The energy loss identifies the atomic number, as shown in figure 6, whilst the flight time gives information on the mass. The position information from the PPACs allows reconstruction of the trajectory. This is illustrated in figure 7, where the X-Y positions of ^{6}He ions are shown in the left panels. The rings in the spectrum from PPAC2, closest to the secondary target, correspond to the discrete ^{6}He energies arising from the production reaction, as seen in figure 6 (top left). By linear extrapolation of the positions registered in the PPACs, the trajectory onto the secondary target is determined, allowing correction for the incidence angle and interaction point of the RIB particle on the target. The closest point of the trajectory to the axis defines the focal length. Measured at low beam intensities, where the Si ΔE-E telescope could be used, the bottom right panel shows the correspondence between the focal length determined from the tracking detectors, and the energy of the ^{6}He as measured in the Si detectors. There is a good correspondence, showing that the focal length information can be used to place cuts on the energy of the radioactive beam. These measurements demonstrate that the thin PPACs, operating at 7 mbar, are able to provide all the information required to define the characteristics of the beam particles passing through them. This now allows measurements of elastic and inelastic scattering, transfer, breakup, fusion and fission in collisions of the light radioactive beams accessible with this device, bombarding any chosen target. Using the large solid angle Si double-sided strip detectors (DSSSDs) illustrated in figure 5, which currently cover a solid angle of π steradians, the beam intensity (several 10^{6}/sec for ^{8}Li, several 10^{5}/sec for ^{6}He) allows a wide range of measurements to be made. A series of measurements is planned for 2016 and onwards.

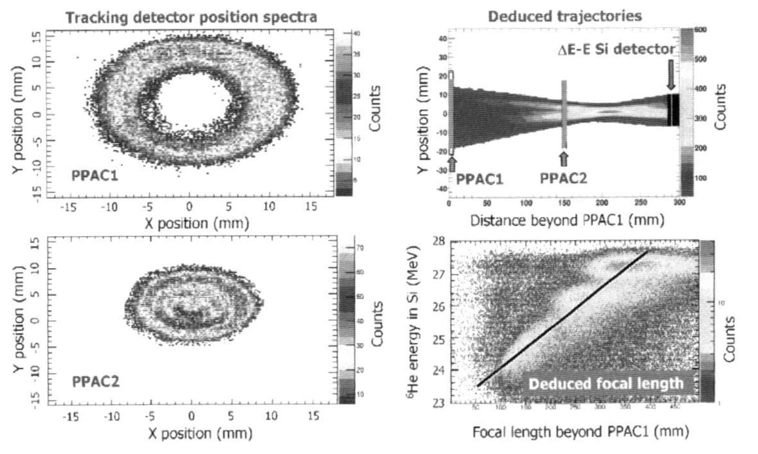

Figure 7. Position spectra for ^{6}He ions measured simultaneously in the two tracking PPACS (left). From them, trajectories are determined (top right), giving the interaction point at the position of the secondary target. The closest distance to the axis is defined as the focal length, which correlates well with the energy measured in the Si detector telescope, showing that identification of all key characteristics is achieved by the tracking detectors

5. Acknowledgments

Support for operations of the ANU Heavy Ion Accelerator Facility through NCRIS is acknowledged. This work was supported by ARC grants DP0879679, FL11010098, DP130101569 and DE140100784.

6. References

[1] Rodriguez M D *et al* 2010 *Nucl. Instr. Meth. in Phys. Res. A* **614** 119.
[2] Leigh, J R *et al* 1995 *Phys. Rev. C* **52** 3151
[3] Dasgupta M *et al* 1998 *Annu. Rev. Nucl. Part. Sci.* **48** 401
[4] Dasgupta M et al 2007 *Phys. Rev. Lett.* **99** 192701
[5] Tanihata I *et al* 1985 *Phys. Lett. B* **160** 380
[6] Hinde D J and Dasgupta M, 2004 *Nature* **431** 748
[7] Rafiei R et al 2011 *Nucl. Instr. Meth. in Phys. Res. A* **631** 12
[8] Horsley A J *et al* 2011 *Nucl. Instr. Meth. in Phys. Res. A* **646** 174

Introducing dynamic dosimaging: potential applications for MRI-linac

P Metcalfe[1], SJ Alnaghy[1,2], M Newall[1], M Gargett[1,4], M. Duncan[1], G Liney[1-4], J Begg[2,3], B Oborn[1-6], M Petasecca[1], M Lerch[1], A Rosenfeld[1]
[1]Centre for Medical Radiation Physics, University of Wollongong, NSW, Australia
[2]Ingham Institute for Applied Medical Research, Sydney, NSW, Australia
[3]Department of Medical Physics, Liverpool and Macarthur Cancer Therapy Centre, Sydney, NSW, Australia
[4]South Western Sydney Clinical School, University of New South Wales, Sydney, NSW, Australia
[5]Royal North Shore Hospital, Sydney, NSW, Australia
[6]Illawarra Cancer Care Centre, Wollongong, NSW, Australia

E-mail: metcalfe@uow.edu.au

Abstract: The new era of intra-fraction dose tracking in radiation therapy delivery demands new dosimetry methods, whereby a moving frame of reference as a function of time may be required. This introduces a new paradigm into radiation therapy dose verification. The term we propose to describe this is *dynamic dosimaging*, which by our definition is tracking the location of a dosimeter array in real time during on-line radiation dose acquisition.

1. Introduction

As described by Ernest Rutherford: "Radiation may be investigated by two methods, one depending upon the action of the photographic plate and the other on the discharge of electrification ... much more rapid than the photographic method and admits of fairly accurate quantitative determination." [1]. We use ancestral versions of this in the form of radiochromic film or 3D gel formulations [2, 3], and ionisation chambers or various semiconductor dosimeters, respectively [4, 5].

A dosimeter can be defined generally as any device that is capable of providing a reading that is a measure of the absorbed dose deposited in its sensitive volume by ionizing radiation [6, 7]. Strictly, radiation dosimetry (or simply "dosimetry") deals with the measurement of the absorbed dose or dose rate resulting from the interaction of ionising radiation with matter [8]. The units of dose in radiotherapy are the Gray (Gy) [9]. We increasingly compare in-beam dose maps with computer-predicted dose maps.

Imaging is the representation or reproduction of an object's form, especially a visual representation (i.e. the formation of an image). We are progressively using dosimetry maps (i.e. images or pictures) of the dose distribution at a dose plane usually, but not exclusively, perpendicular to beam delivery. Just as we call X-ray and CT anatomical information images, we propose that we start calling a dose map a 'dose-mage', or following the dosimeter convention and instead dropping the 'e', i.e. a 'dosimage'. The method of acquiring processing and reproducing these images over time or over a snap shot in time would be given the name 'dosimaging'.

Examples of typical dosimage acquisitions would be an Electronic Portal Imaging Device (EPID) or dosimeter array acquisition of a pre-patient IMRT delivery. Different representations of dosimage display would include dose as a colour wash, isodose map or pixelated dose map (grey scale or colour).

As radiotherapy technology improves, we are witnessing advanced techniques of delivery. Static gated delivery is the first step beyond static delivery. The most elegant and time efficient systems involve dynamic tracking dose delivery systems, which follow the tumour motion and hence encapsulate the tumour with dose. This is called intra-fraction dose tracking or dynamic dose delivery. It enables the dose margin around the tumour to be reduced with an associated small treatment volume. Hence less normal tissue is treated with an associated reduction in normal tissue toxicity.

There are examples of existing technology that provide scenarios where the patient's tumour moves and is tracked dynamically by the treatment device. These include CyberKnife® or multileaf collimator (MLC) tracking with Calypso® RF tracking device [10].

As with current radiation therapy quality assurance (QA), which validates static dose delivery, we need a mechanism to validate dynamic tracking dose delivery. In order to best estimate the dose to the target using a surrogate phantom dosimeter delivery, there are two choices:

(1) Leave the detector static and re-compute the dose distribution using a movement that is simulated in a virtual reality environment (we have not tested this here).
(2) Move the detector array in the same temporal and spatial pattern that the tumour would move.

This introduces a new paradigm into radiation therapy dose verification. The term we propose for moving the detector in the tumour pattern while acquiring dose is 'dynamic dosimaging'. This describes tracking the location of a dosimeter array in real time during radiation acquisition. This is important for dynamically moving dose deliveries. In essence, the dosimeter is moving in time in the same frame of reference as that of the tumour motion. A summary defining the terms dosimaging, static dosimaging, dynamic dosimaging, and virtual dynamic dosimaging is outlined in Figure 1.

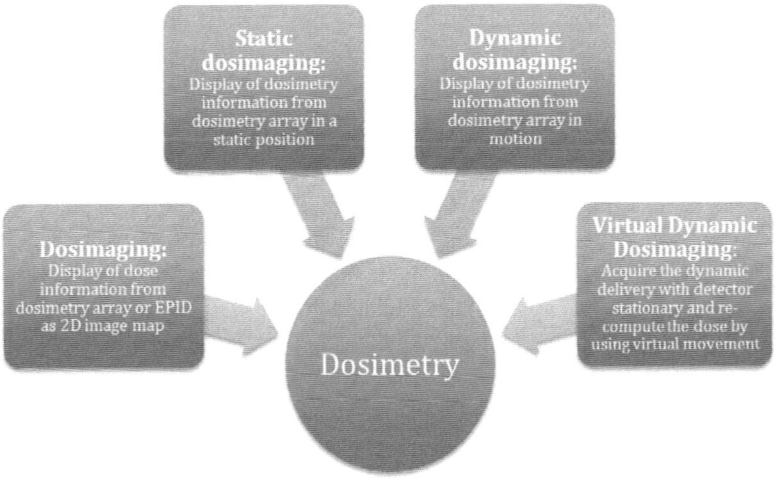

Figure 1. Summary of different definitions of dosimaging as subsets of dosimetry

Note that the dose map acquisition with the dosimeter array in a static fixed position (i.e. static dosimaging) is extremely useful for the majority of radiotherapy dose deliveries. These include IMRT, VMAT, SRT and SBRT/SABR treatments and gated deliveries. Hence leaving the dosimeter array in a static location may remain preferable perhaps for a majority of QA dosimetry. Subsequently, less normal tissue is treated. Then, why move the dosimeter array? By moving the phantom on a platform using a typical patient breathing trace, this ensures acquisition of the ground truth dose, provided that all components of system lag (e.g. MLC tracking delay) are accounted for.

2. Methods

Whilst not so far named, we have previously demonstrated in the literature an example of dynamic dosimaging using a high resolution array dosimeter, the MagicPlate-512 (M512). This dosimeter is combined with a respiratory motion hexapod phantom that mimics and tracks lung motion during radiation acquisition in RF fields using Calypso RF tracking device [10]. The method is not necessarily restricted to this technology and there may be other devices that can or have already achieved results resembling this approach.

The essential components for a generic dynamic dosimaging system are:

(1) Platform that can move synchronously with a patient movement (usually breathing or specific individual organ motion) pattern.
(2) Array dosimeter that can be placed in-phantom on the platform
(3) A platform-dosimeter system that can mimic motion by tracking the position of tumour motion in time and space using fiducials or another tracking mechanism.
(4) Software that can record the dose with high temporal resolution in real time with respect to detector position.

Because the frame of reference of the beam's eye view is now a moving position in space and time, then the enemy of dynamic dosimaging is system latency. If the latency in time is significant compared to the movement cycle then anomalies can occur in the acquired dose distribution.

3. Results

Figure 2 presents an example of a typical static dosimage that was collected using an EPID acquisition and processing software described elsewhere [11].

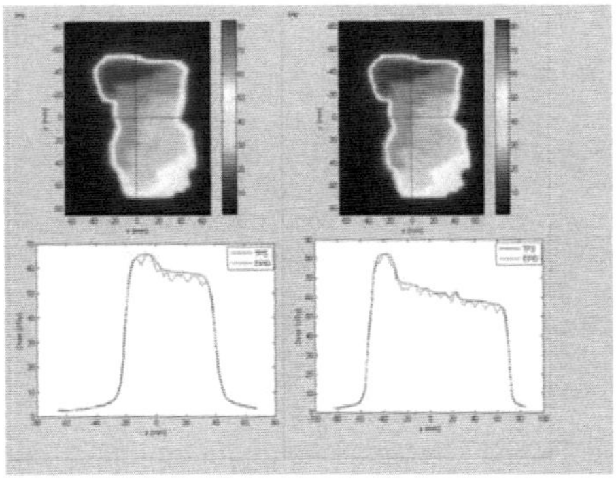

Figure 2. Static dosimage acquired using an Electronic Portal Imaging Device (EPID): Treatment planning system (TPS) dosimage (top left) and profile (bottom left) and EPID reconstructed dosimage (top right) and dose profile (bottom right).

To give a qualitative example of what can be achieved using dynamic dosimaging, Figure 3 shows a pixelated colour map of three different delivery and acquisition combinations. These detectors have been successfully placed in respiratory motion phantoms for mimicking lung motion during radiation acquisition in RF fields using Calypso RF tracking device [12].

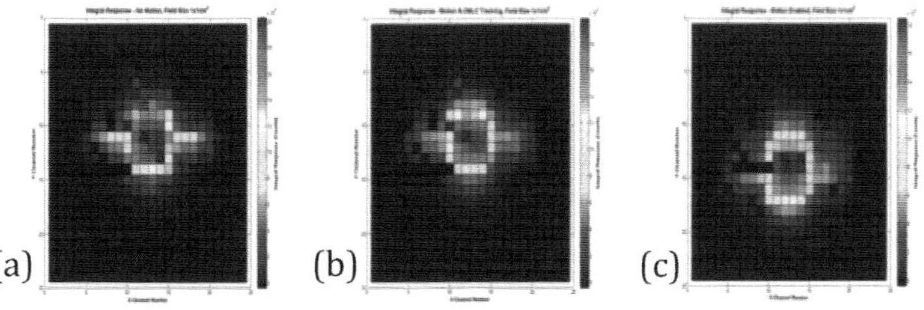

Figure 3. M512 array acquisitions, (1 × 1cm MLC) (a) static MLC, static dosimeter, (b) static MLC, moving platform, (c) moving MLC moving platform (i.e. dynamic dosimaging).

4. Discussion and Work in Progress

There are other potential methods to track the detector positions. These may include using x-ray imaging combined with tracking methods to locate the detector in time and space. These could be either using stereoscopic tracking methods such as ExacTrack® [13] or respiratory correlated cone beam CT (CBCT) [14].

We envisage this as a potential future QA method for MRI-linac dynamic tracking deliveries. Solid objects in MRI (e.g. bone, ultra-short relaxation time) are inherently difficult to visualise. In this case, one option is to place MRI fiducials on a detector array to indicate its position in space (see Figure 4). We have obtained this image using a 3 T MRI scanner. We have achieved moving and visualising the detector during a MRI acquisition.

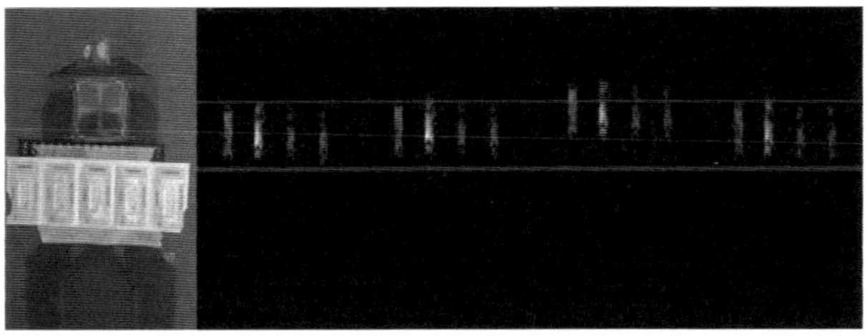

Figure 4. (a) Octa diode array with five MRI fiducial markers, (b) Fiducials tracked using MRI breathing sequence showing movement shown (i.e. 1, 2, 3, 4).

5. Conclusion

The imperative for moving a dosimeter during pre-treatment verification follows the trend toward dynamic tumour tracking. To validate dose to a moving tumour via a dynamic delivery platform or, *dynamic dosimaging*, is one emerging technique that will satisfy the requirement that the dose you see (predicted by the TPS) is the dose you get.

6. References

[1] Rutherford E 1899 Ed. J Chadwick, RTF Publishers, London, 1st Ed 1962, 2nd Ed2014
[2] Baldock C *et al Phys. Med. Biol.* 55 R1-63
[3] Murry P *et al* 2000 *Australas. Phys. Eng. Sci. Med.* **23** 44-51
[4] Vial P *et al* 2008 *Med. Phys.* **35** 1267-77
[5] Greer P B *et al* 2007 *Med. Phys.* **34** 4389-98
[6] Hill R *et al* 2010 *Med. Phys.* **37** 4355-63
[7] Hill R *et al* 2009 *Med. Phys.* **36** 3971-81

[8] Attix F 1986 Wiley VCH 264
[9] Cameron J 1991 Radiation Dosimetry *Environmental Health Protection* **91** 45-48
[10] Keall P *et al* 2014 *Med. Phys.* **41** 020702
[11] Xing A *et al* 2014 *J. Phys.: Conf. Series* **489** 012084
[12] Petasecca M *et al* 2015 *Med. Phys.* ***42*** 2992
[13] Sonke J J *et al* 2005 *Med. Phys.* **32** 1176
[14] Lee S W *et al* 2008 *J. Appl. Clin. Med. Phys.* **9** 2318

Mitochondria as a target for radiosensitisation by gold nanoparticles

S J McMahon[1-3], A L McNamara[1,3], J Schuemann[1], K M Prise[2], H Paganetti[1]
[1]Department of Radiation Oncology, Massachusetts General Hospital & Harvard Medical School, Boston, USA
[2]Centre for Cancer Research and Cell Biology, Queen's University Belfast, Belfast, UK

E-mail: stephen.mcmahon@qub.ac.uk

Abstract. While Gold Nanoparticles (GNPs) have been extensively studied as radiosensitisers in recent years, there is a lack of studies of their impact on targets outside of the cell's nuclear DNA. We present Monte Carlo simulations of the energy deposited by X-ray irradiation in mitochondria in cells with and without cytoplasmic GNPs. These simulations show that the presence of GNPs within the cytoplasm can significantly increase (3-4 fold) the number of ionisation clusters of both small and large sizes. As these clusters are strongly associated with DNA damage, these results suggest that mitochondrial DNA may be a significant target for GNP radiosensitisation when the nanoparticles cannot penetrate the cell nucleus.

1. Introduction

For more than a decade, Gold Nanoparticles (GNPs) have been the subject of extensive investigations as radiosensitising agents, following landmark work by Hainfeld *et al* [1]. By combining gold's high atomic number and perceived biocompatibility with the ability of sub-100 nm objects to penetrate the tumour vasculature, it was believed that GNPs acted effectively to increase dose to target structures. While there is now extensive evidence that GNPs radiosensitise cells, it is increasingly clear that their mechanism is not simply dose enhancement as originally believed. Experimental observations of GNP radiosensitisation have shown significantly greater effects than would be predicted from physical dose enhancement alone, suggesting that other processes drive radiosensitisation by GNPs [2].

In addition to the chemical and biological stresses which may be introduced by nanoparticles in a cellular system, there has also been interest in better understanding the physical distribution of dose on the sub-cellular scale. Early nanoscale models of GNPs demonstrated that interactions between radiation and GNPs led to dramatic dose heterogeneity on length scales of hundreds of nanometres, with localised doses of hundreds to thousands of Gray being deposited within volumes of tens to hundreds of nanometres in diameter. Analysis of these highly heterogeneous dose distributions through techniques developed for heavy ion therapy, such as the Local Effect Model (LEM), indicated that these dose heterogeneities drive significantly greater sensitisation than predicted from macroscopic dose enhancement alone, potentially explaining the increased radiosensitisation[3].

There remain significant questions about these early analyses, however, as they typically assumed uniform distribution of GNPs throughout a sensitive target volume, generally assumed to be the DNA contained within the cell nucleus. By contrast, in experimental systems many GNP preparations are seen

[3] These authors contributed equally to this work

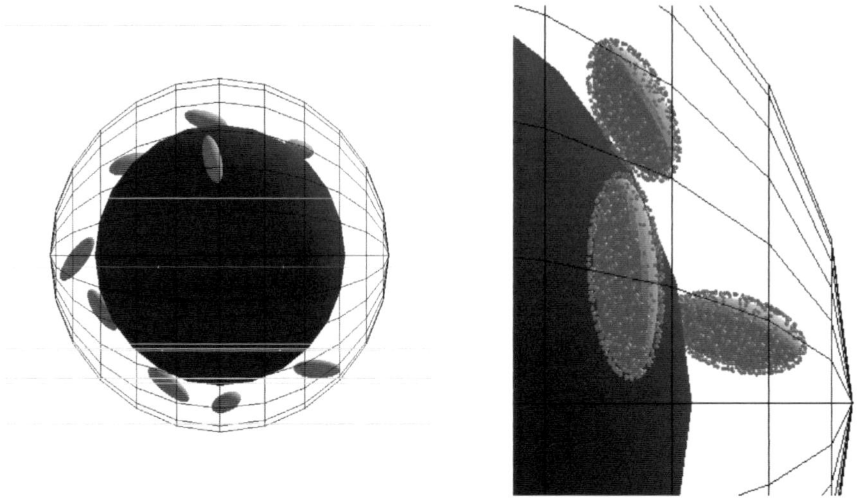

Figure 1. Left: Schematic illustration of modelled cell volume, showing outer cellular membrane, nucleus (blue) and smaller mitochondria (green). Right: Close-up of mitochondria, showing distribution of surface-bound GNPs (yellow).

to reside primarily or exclusively within the cytoplasm. In these cases, GNPs are necessarily tens to hundreds of nanometres away from any nuclear DNA, dramatically reducing the degree of dose heterogeneity seen by the DNA and the sensitisation predicted by models such as the LEM [4]. As significant sensitisation is seen in many such systems, this then raises the question of other cellular targets which may drive radiosensitivity.

One significant extra-nuclear target is the mitochondrion. These organelles are essential for cellular function (e.g. metabolism, apoptosis regulation, signalling, reactive species production). Mitochondria are the only extra-nuclear site to contain genetic material, but lack functional DNA repair [5], suggesting they may be sensitive to highly localised dose distributions in a similar fashion to the nucleus [6]. Importantly, as mitochondria are distributed throughout the cytoplasm and are small compared to the nucleus, mitochondrial DNA is much more likely to be close to a GNP and thus exposed to a high dose. Furthermore there is evidence that suggests GNPs may naturally accumulate on the surface of the mitochondria due to surface charge properties of the membrane [7]. To date, however, there have been few detailed studies of the dose and ionisation distributions within mitochondria when cells are irradiated in the presence of GNPs. In this work, we present an analysis of ionisation clustering effects in mitochondria with and without GNPs, to assess the potential significance of these effects in cellular radiosensitisation.

2. Methods

Radiation interactions with cells were calculated using the Geant4 Monte Carlo toolkit (version 10.1.p02) [8], including the Geant4-DNA package [9,10] to simulate ionisation on an event-by-event basis on the nanometre scale. A simple cellular model was implemented, consisting of an 11.5 μm diameter spherical cell, which contains an 8.5 μm diameter nucleus, and 30 elliptical mitochondria, with dimensions of 1.8 μm × 1 μm × 0.6 μm along their axes, shown in Figure 1. These mitochondria were distributed randomly throughout the cytoplasm with random orientation. Exposure to 40 nm diameter GNPs was modelled, with 50 GNPs distributed at the surface of each mitochondrion.

Cells were exposed to monoenergetic X-rays of either 50 or 100 keV, with a total of 3.0×10^8 incident primary photons. The distribution of energy deposited within the mitochondria and nucleus was scored, together with the positions of ionisation and excitation events. To sample ionisation cluster sizes, virtual 'DNA segments' were placed within the irradiated volume. These were 10 nm by 5 nm cylinders, corresponding to approximately a 20 base pair segment of DNA. By scoring the density of ionisations

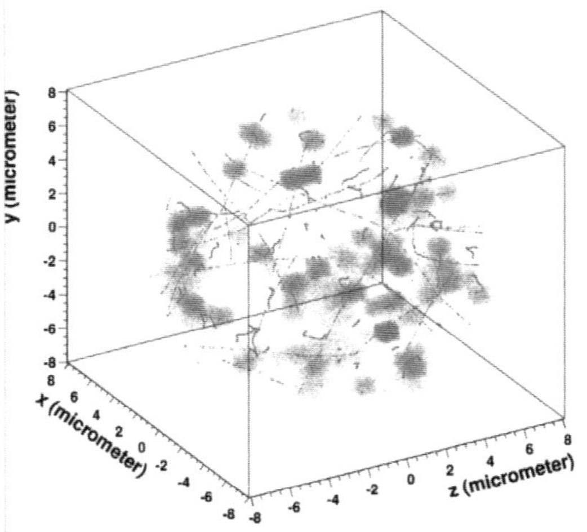

Figure 2. Map of secondary electron tracks for a simulated broad beam of 50 keV incident photons within the entire cell volume, 10^6 photons were modelled. Spontaneous clustering of ionisation events can be seen throughout the volume.

in these volumes, an estimate of the degree of ionisation clustering, and thus DNA damage, can be obtained [11].

3. Results

Figure 2 shows an illustration of the distribution of ionising events from a broad beam of 50 keV photons in the whole cell volume. Even in the absence of GNPs, stochastic clustering of ionisation events can be observed. Figure 3 shows the number of ionisation clusters of various sizes induced in mitochondria exposed to 50 or 100 keV X-rays, either with or without GNPs as described above. Although the total dose increase within the cell may be relatively small in these conditions, the number of ionising events increases substantially – by a factor of 3 or more. These effects are seen across all ionisation cluster sizes from 1 to 20 ionisation/excitation events per DNA segment, and for both 50 and 100 keV incident X-rays. By contrast, the impact on ionisation clustering for points within the nucleus is relatively small (a factor of 1.5 or less), indicating that DNA is subject to significantly more sensitisation in mitochondria rather than within the nuclear volume.

4. Conclusions

Cytoplasm-bound GNPs are not expected to lead to large increases in ionisation density within the nucleus, and thus are expected to lead to only modest sensitisation of nuclear DNA. However the small size of mitochondria, combined with their close proximity to cytoplasmic GNPs suggests that mitochondrial DNA will be subject to significant additional damage – with a several-fold increase in the number of ionisation clusters of all sizes in this simulation. If such clusters translate into DNA damage as is seen in the nucleus, then this may lead to a significant increase in mitochondrial damage and corresponding sensitisation of these organelles. Disruption of mitochondrial function can in turn have significant impact on the general functions of the cell. Mitochondria are essential for cellular metabolism, and disruption or dis-regulation of their function can lead to significant alterations in cellular energy transport, replication, and levels of oxidative stress within the cell. All of these effects may in turn impact on the cell's ability to survive exposure to ionising radiation, either by directly causing sufficient disruption of cellular function to lead to cell death or by combining with direct damage to nuclear DNA to prevent recovery following radiation exposure [6].

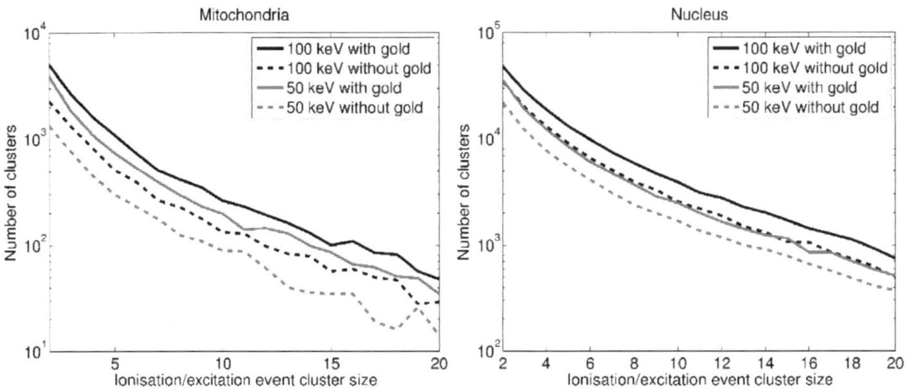

Figure 3. Impact of GNPs on ionisation clustering in mitochondria (left) and nuclei (right) for cells exposed to 50 keV (red) or 100 keV (black) X-rays. While ionisation rates increase in both regions, this is more pronounced in mitochondria (2.5-3 fold) than the nucleus (~1.5 fold).

These results indicate that, in cases where GNP are not internalised to the cell nucleus, mitochondria may be an important target for radiosensitisation through damage of mitochondrial DNA and disruption of the cellular metabolism. Further simulations to quantify the biological impacts of this ionisation clustering on mitochondrial DNA breakage, coupled with experimental investigations of disruption of mitochondrial function in the presence of GNPs, are needed to more accurately quantify the contribution of this pathway to GNP radiosensitisation.

5. Acknowledgements

The authors would like to thank the following funders for supporting their work: SJM - EC FP7 (grant MC-IOF-623630); KMP - EC FP7 (grant MC-ITN-608163); ALM, HP and JS NIH/NCI (grant R01 CA187003).

6. References

[1] Hainfeld J F et al 2004. *Phys. Med. Biol.* **49** N309-15
[2] Butterworth K T et al 2012. *Nanoscale* **4** 4830-8
[3] McMahon S J et al 2011 *Sci. Rep.* **1** 18
[4] Lin Y *et al* 2015 *Phys. Med. Biol.* **60** 4149-68
[5] Larsen N B *et al* 2005 *Mitochondrion* **5** 89-108
[6] Kam W W-Y and Banati R B 2013 *Free Radic. Biol. Med.* **65** 607-19
[7] Karataş Ö F *et al* 2009 *Colloids Surfaces B Biointerfaces* **71** 315-8
[8] Agostinelli S *et al* 2003 *Nucl. Instruments Methods Phys. Res. Sect. A* **506** 250-303
[9] Incerti S *et al* 2010 *Med. Phys.* **37** 4692-708
[10] Bernal M A *et al* 2015 *Phys. Medica* **31** 1-14
[11] Nikjoo H *et al* 1989 *Phys. Med. Biol.* **34** 691-705

IOP Conf. Series: Journal of Physics: Conf. Series **777** (2017) 012009 doi:10.1088/1742-6596/777/1/012009

New 3D Silicon detectors for dosimetry in Microbeam Radiation Therapy

M L F Lerch[1], A Dipuglia[1], M Cameron[1], P Fournier[1], J Davis[1], M Petasecca[1], I Cornelius[1], V Perevertaylo[2], A B Rosenfeld[1]
[1]Centre for Medical Radiation Physics, University of Wollongong, NSW, Australia, 2522
[2]SPA-BIT, Kiev, Ukraine

Email: mlerch@uow.edu.au

Abstract. Microbeam Radiation Therapy (MRT) involves the use of a spatially fractionated beam of synchrotron generated X-rays to treat tumours. MRT treatment is delivered via an array of high dose 'peaks' separated by low dose 'valleys'. A good Peak to Valley Dose Ratio (PVDR) is an important indicator of successful treatment outcomes. MRT dosimetry requires a radiation hard detector with high spatial resolution, large dynamic range, which is ideally real-time and tissue equivalent. We have developed a Silicon Strip Detector (SSD) and very recently, a new 3D MESA SSD to meet the very stringent requirements of MRT dosimetry. We have compared these detectors through the characterisation of the MRT radiation field at the Australian Synchrotron Imaging and Medical Beamline. The EPI SSD was able to measure the microbeam profiles and PVDRs, however the effective spatial resolution was limited by the detector alignment options available at the time. The geometry of the new 3D MESA SSD is less sensitive to this alignment restriction was able to measure the microbeam profiles within 2μm of that expected. The 3D MESA SSD measured PVDRs were possibly affected by undesired and slow charge collection outside the sensitive volume and additional scattering from the device substrate.

1. Introduction
Microbeam Radiation Therapy (MRT) is a preclinical radiotherapy modality characterised by use of a synchrotron light source to produce highly collimated, micron-scale striated x-ray fields under development to treat radiation resistant and otherwise untreatable tumours. Typical MRT configurations consist of multiple high dose 'peaks' of 25-75 μm separated by low dose 'valleys' 100-400 μm wide and are produced by a spectrum of X-rays in the range of 50-600 keV with treatment doses that can be delivered via dose rates of up to 20 kGy/s. The high dose rate (>1 kGy/s) and low divergent synchrotron X-rays are required for MRT in order to ensure the delivery of the spatially fractionated treatment dose profile to the tumour site. The Peak to Valley Dose Ratio (PVDR) and geometrical configuration of the microbeams is one of the main factors that are believed to be behind the success of MRT that leads to the minimisation of normal tissue damage while inflicting significant damage to the tumour tissue. A detector suitable for dosimetry of MRT must address the issues of very high dose rate and dose gradients that exist within the striated treatment field. This requires a dosimeter with micron size spatial resolution to resolve individual X-ray microbeams, and a large dynamic range (approximately 10^5) in order to simultaneously resolve instantaneous dose rate in both the peak and the valley. While conforming to both of these criteria, the dosimetry system must also be able to evaluate all of the main beam parameters quickly (ideally in real-time) for fast quality assurance of the treatment plan immediately prior to MRT treatment.

IOP Conf. Series: Journal of Physics: Conf. Series **777** (2017) 012009 doi:10.1088/1742-6596/777/1/012009

The Centre for Medical Radiation Physics (CMRP) has developed the X-Tream dosimetry system for use in quality assurance of MRT [1]. The system has a sampling rate of 1 μs and is coupled to a silicon strip detector (SSD). A variety of SSDs have been produced to address the challenges of high spatial resolution, high dynamic range, and real time response. The use of silicon detectors in an edge-on orientation [2] (rather than face-on where the face is perpendicular to the beam) has been investigated by CMRP as a way to increase spatial resolution - as in this geometry the spatial resolution is determined mainly by the thickness of the detector as opposed to the cross-sectional width when face-on. However, this introduces a non-water equivalent response due to interaction in the silicon surrounding the Sensitive Volume (SV) and substrate backing. The 3D MESA SSD was developed as a first step in overcoming some of the disadvantages of silicon dosimeters such as their energy dependence via selective etching of the silicon surrounding the sensitive volume (SV). The system and two generations of SSD (EPI and 3D MESA) were tested and characterised at the Australian Synchrotron (AS) on the Imaging and Medical Beam Line (IMBL) in hutch 1B. This hutch has installed a multislit collimator that produces horizontal microbeams of width 25 μm or 50 μm and pitch of 200 μm (denoted as '25/200' and '50/200' respectively).

2. Materials and Methods

The EPI SSD is a 10 μm wide silicon diode fabricated on a 50 μm thick epitaxial substrate, and then grown on top of a 370 μm thick p-type substrate, schematic shown in Figure 1. The structure itself incorporates an n+ guard ring which limits the SV of the device to approximately 30 μm, a detailed description of the device can be found in [3].

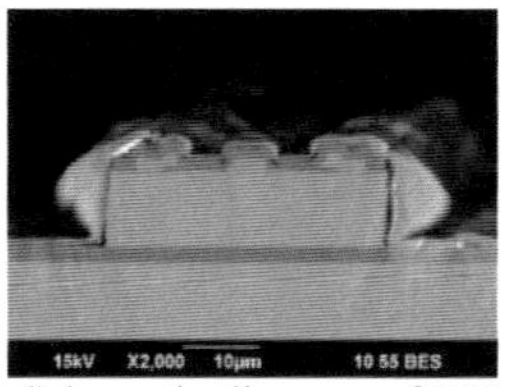

Figure 1. Schematic diagram of EPI SSD [3].

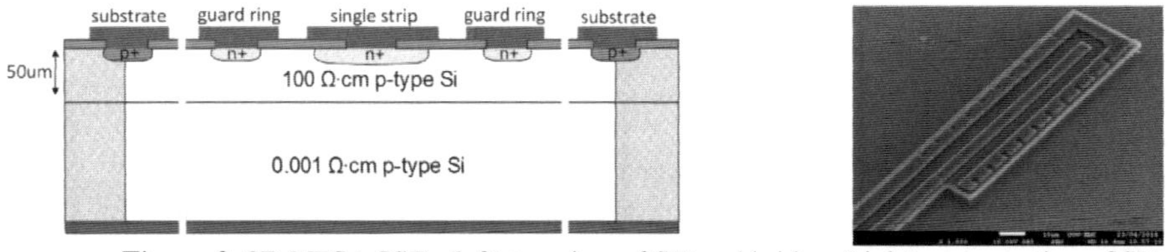

Figure 2. 3D MESA SSD: (left) top view of SV and bridge; (right) cross section of SV.

The 3D MESA SSD is an n-type Silicon-on-Insulator (SOI) device on a 300 μm thick silicon substrate with a SV 36 μm x 250 μm that has had surrounding silicon etched away 10 μm deep to the SiO_2 insulating layer. The middle contact is p+ doped and surrounded by a continuous n+ doped contact. Electron microscope images of the 3D MESA taken at the Australian Institute of Innovative Materials (AIIM) are shown in Figure 2. The 3D MESA and EPI devices were characterised at CMRP and tested in an MRT irradiation field at the AS, with readout of the devices performed by the X-*Tream* System. An operational reverse bias voltage of 60 V and 5 V for the EPI [3] and 3D MESA devices respectively was used. The filter configuration used at the AS is shown in Table 1 and experimental setup shown in Figure 3.

Before characterisation of the MRT treatment field profile, both devices were subjected to pre-irradiation in order to stabilise their response with irradiation that was performed in broadbeam mode (where the Multi-Slit Collimator (MSC) is removed) for a 2 x 10 mm² field. The EPI SSD was placed at the surface of a solid water (RMI-457) phantom (10 x 10 x 10 cm³) in face-on orientation and exposed for a five second exposure. The same procedure was performed for the 3D MESA SSD but for only 200 ms exposure was required to demonstrate the stability of its response.

The SSDs were then calibrated at 2 cm depth in the solid water phantom in a 2x2 cm^2 uniform X-ray field (same photon energy spectrum as MRT), via a PTW Pinpoint^{TM} ion chamber (model 31015). Characterisation of the MRT profile was then performed with the MSC in place by the EPI SSD and 3D MESA in edge-on orientation at 2 cm depth. Profiles were obtained by scanning the SSDs through the intrinsic MRT field containing 5 horizontal microbeams. The two microbeam geometries used in this experiment were 25/200 and 50/200. The measured FWHM of the microbeams as well as the PVDRs were then able to be deduced from the profiles.

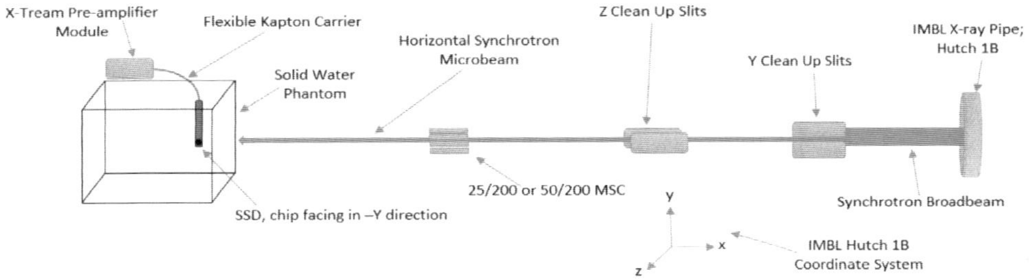

Figure 3. Experimental setup of experiments performed at AS IMBL, hutch 1B.

Table 1. Filter configuration used at AS.

Paddle	#1	#2	#3	#4	#5
Material	C	C[HD]	C[HD]	Cu	Cu
Thickness	4.5 mm	5 mm	10 mm	1 mm	1 mm
Angle	90°	45°	45°	45°	45°

3. Results

Figure 4 shows the pre-irradiation of the (A) EPI and (B) 3D MESA SSDs. Figure 5 shows the 50/200 MRT intrinsic field profiles measured by the (A) EPI and (B) 3D MESA SSDs. The PVDRs and FWHM measured by the detectors for both beam geometries are shown in Table 2.

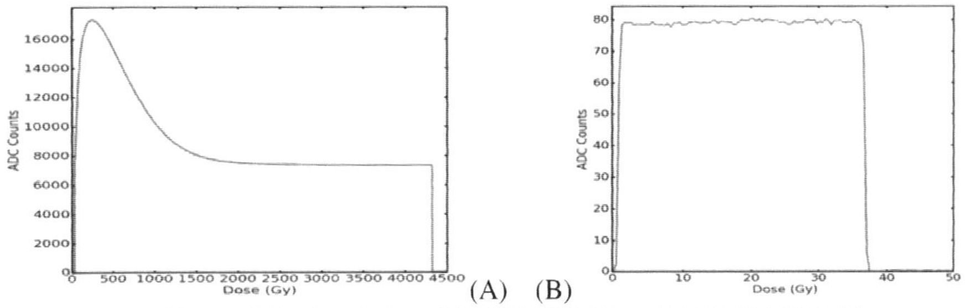

Figure 4. Pre-irradiation of the EPI SSD (A) and 3D MESA SSD (B).

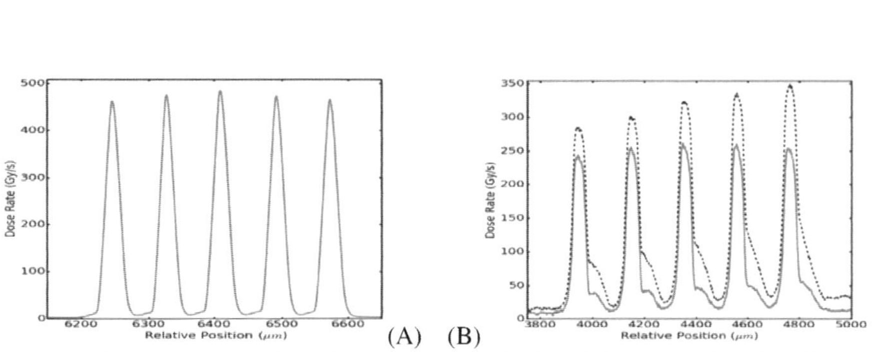

Figure 5. Measured 50/200 MRT profile using the (A) EPI SSD and (B) 3D MESA SSD showing initial scan in solid red and subsequent scan in dotted black (B).

Table 2. Comparison of FWHM and PVDR for the 3D MESA and EPI SSDs.

	EPI		3D MESA	
Beam geometry	PVDR	FWHM (μm)	PVDR	FWHM (μm)
25/200	41±3	45±4	35±4	26±1
50/200	38±2	56±5	14±2	52±1

4. Discussion

The significant change in EPI SSD response with total irradiation dose is due to the reduction in the undesired charged collection from under the wire bonding pads. Beyond 2.5 kGy of pre-irradiation the response is stable (reduces by 1% per kGy). For the 3D MESA SSD, the response is stable immediately upon irradiation. Minimising the surrounding silicon from the bonding pads and sensitive volume appears to make the pre-irradiation of the 3D MESA redundant. The response of the 3D MESA is ~100 times less than the EPI due to its much smaller SV.

The EPI SSD measured larger FWHMs than expected (especially for the 25/200 MSC) due to volume averaging effects over the larger effective SV resulting from the detector misalignment with respect to the microbeams (see Figure 5A). The result clearly highlights one of the challenges in MRT dosimetry. On the other hand, the 3D MESA measured FWHMs within ±2 μm expected value defined by the microbeam defining slots in the MSC (see Table 2).

The PVDRs for the 25/200 MSC geometry agreed for both SSDs. For the 50/200 MSC geometry however, the EPI SSD measured PVDRs that were significantly larger than that measured by 3D MESA. PVDRs are poor for 3D MESA due to the structure observed in the valley regions. Further investigation shows that the 3D MESA profiles were affected in two ways. The edge-on orientation led to small 'bumps', to the left of the peak profiles (see Figure 5B) which is the 'substrate' side of the SSD. It is possible that these due to a contribution from electrons generated in the substrate as previously reported by Rosenfeld [2]. In addition, repeated profile scans produced an increase in baseline during these consecutive scans which we attribute to very slow charge collection generated under the bridge between the active volume and the wire bonding pads. This effect was subsequently investigated using Ion Beam Induced Charge Collection (IBICC) studies (not shown), which indicated determined that this effect did not occur at reverse biases less than 3 V. The experimental methodology will therefore be adjusted for future experiments to address these issues.

5. Conclusion

Two generations of single silicon strip detectors (EPI SSD and 3D SOI MESA) have been developed at the Centre for Medical Radiation Physics, University of Wollongong, for real-time MRT dosimetry. The SSDs were compared by characterising the 25/200 and 50/200 intrinsic MRT fields at the Australian Synchrotron. The geometry of the new 3D MESA SSD was found to be less sensitive to mechanical alignment restrictions and was clearly able to measure the FWHM of the X-ray microbeam profiles however features observed in the valley affected the PVDR estimation in the MRT field. Future work will include investigation into reducing thickness of the silicon substrate in edge-on orientation and examination of the effect of bias on the observed valley features and PVDRs.

6. Acknowlegements
The authors wish to thank AIIM, UOW and the Australian Synchrotron IMBL team for their excellent assistance. The authors wish to acknowledge the financial support of the NH&MRC (Development Grant APP1093256) and the Australian Synchrotron (Beamtime Grant AS152/IM/9404 and AS153/IM/10045).

7. References
[1] Lerch M *et al* 2011. *Radiat. Meas.* **46** 1560-65
[2] Rosenfeld A *et al* 2005. *IEEE Trans. Nucl. Sci.* **520** 2562-69
[3] Petasecca M *et al* 2012. *J.Instrum* **7** 1-15

Hybrid approach to calculating proton stopping power in hydrogen

J J Bailey[1], **A S Kadyrov, I B Abdurakhmanov, I Bray**
Curtin Institute for Computation and Department of Physics, Astronomy and Medical Radiation Sciences, Curtin University, GPO Box U1987, Perth 6845, Australia

E-mail: jackson.bailey@postgrad.curtin.edu.au

Abstract. Proton stopping power in hydrogen is calculated using a hybrid method. A two- centre convergent close-coupling method is used for calculations involving the proton fraction of the beam, while the Born approximation is used for the hydrogen fraction. For proton-hydrogen collisions rearrangement processes are explicitly included via a two-centre expansion. Hydrogen-hydrogen collisions are calculated including one- and two-electron processes. Despite using the first-order approximation in the hydrogen-hydrogen channel, overall reasonably good agreement with experiment is seen above 100 keV.

1. Introduction

Detailed knowledge of proton beam interactions with atoms and molecules provides a basis for accurate simulations using the low-energy models of Geant4 [1]. The first study of proton stopping in hydrogen was performed by Dalgarno and Griffing [2]. They applied the first Born approximation (FBA) to calculate the proton-hydrogen (p-H) and hydrogen-hydrogen (H-H) stopping powers and combined the results by weighting each contribution by its charge-state fraction. Rearrangement processes in the case of p-H scattering and two-electron processes in the case of H-H scattering were included and agreement with proton-H_2 experiment above 120 keV was obtained. Their calculations underestimated experiment at low energies. This discrepancy was attributed to the failure of the Bragg summation rule since the proton fraction of the beam should be higher for H_2 than for H. Schiwietz [3] performed single-centre coupled-channel atomic-orbital (AO) calculations for the proton fraction of the beam. They used FBA calculations for the hydrogen fraction (including single excitation and single ionisation processes only) and experimental charge-state fractions of Allison [4] to obtain the total stopping power. Agreement with experiment within 5% was obtained at low and high energies however results underestimated the experiment by 10-15% at intermediate energies. It was suggested that the deterioration was due to an inaccurate ionisation cross section in H-H collisions as electron-electron correlations were neglected. Schiwietz and Grande [5] later built on this work by replacing single-centre AO results below 30 keV with two-centre (AO+) ones. They included partial excitation, electron capture, and total ionisation cross sections. Additionally, a hydrogen-like screened potential was used to perform AO calculations for H+H collisions. Continuing to use the experimental H_2 charge-state fractions, these authors achieved 5% agreement with the H_2 stopping-power experiments over the whole energy range. Fainstein *et al.* [6] used the continuum-distorted-wave eikonal-initial-state (CDW-EIS) model to calculate the stopping power of protons impinging on atomic hydrogen. When combined with the FBA H-H results of Dalgarno

[1] To whom any correspondence should be addressed.

and Griffing [2] good agreement with experiment was obtained above 70 keV. Disagreement with the experiment below 70 keV was attributed to the usage of the FBA in the H-H channel.

The convergent close-coupling (CCC) method has recently been applied to calculate stopping powers of antiprotons in atomic [7] and molecular [8] targets. The CCC method is based on expansion of the scattering wave function using a large basis of specially-generated negative- and positive-energy pseudostates. Here we present our first results for the proton stopping power in hydrogen including electron-capture channels.

2. Theory

The stopping power, the energy loss per unit path length, is defined as a product of the stopping cross section S and the number of target atoms per unit volume. Since the proton-hydrogen collision system involves rearrangement processes it is necessary to take into account the charge state of the projectile. The total stopping cross section for the proton-hydrogen system at a given incident energy of the projectile is given by

$$S = f^{\mathrm{H}^+} S^{\mathrm{H}^+} + f^{\mathrm{H}^0} S^{\mathrm{H}^0}, \tag{1}$$

where f^{H^+} (f^{H^0}) is the positive (neutral) charge-state fraction of the beam and S^{H^+} (S^{H^0}) is the stopping cross section for the case of the beam consisting entirely of positively-charged (neutral) projectiles. We neglect the negative charge state as the probability of H^- formation is insignificant. The charge-state fractions can be calculated from the electron-capture and electron-loss cross sections in p-H and H-H collisions. They are given by $f^{\mathrm{H}^+} = \sigma_1/(\sigma_\mathrm{c}+\sigma_1)$ and $f^{\mathrm{H}^0} = \sigma_\mathrm{c}/(\sigma_\mathrm{c}+\sigma_1)$, where σ_c (σ_1) is the electron-capture (electron-loss) cross section. This highlights the importance of full two-centre calculations.

For the positive charge-state contribution we use the two-centre CCC method [9, 10]. Both bound and continuum spectra are modelled via a pseudostate expansion on the projectile and target centres. This type of expansion allows us to accurately calculate rearrangement processes. Since the continuum is discretised we may write the positive charge-state electronic stopping power as

$$S_\mathrm{e}^{\mathrm{H}^+} = \sum_f (\varepsilon_f - \varepsilon_i)\sigma_{fi} + \sum_k (\varepsilon_k - \varepsilon_i + v^2/2)\sigma_{ki}. \tag{2}$$

The first summation represents direct transitions and goes over all the target pseudostates. The second term represents capture processes, both to bound and continuum states depending on whether ε_k is negative or positive. The $v^2/2$ term represents the kinetic energy of an electron travelling with the speed of the incident proton after being captured.

For the neutral charge-state contribution we apply the Born approximation. With this method we are able to model both single-electron (single excitation and single ionisation) and two-electron (double excitation, double ionisation and ionisation with excitation) processes. For bound states we use the known hydrogen wave functions and for the continuum we use a two-body Coulomb wave function assuming that the ejected electron moves in the field of its parent nucleus only. The neutral charge-state stopping cross section is then given by

$$\begin{aligned} S_\mathrm{e}^{\mathrm{H}^0} = {} & 2\sum_f (\varepsilon_f - \varepsilon_i)\sigma_{fi} + 2\int_0^{E_0+\varepsilon_i} (\varepsilon - \varepsilon_i)\frac{d\sigma}{d\varepsilon}d\varepsilon + \sum_{f\neq i}\sum_{k\neq i}(\varepsilon_f - \varepsilon_i + \varepsilon_k - \varepsilon_i)\sigma_{fi,ki} \\ & + \int_0^{E_0+\varepsilon_i}\int_0^{E_0+\varepsilon_i}(\varepsilon - \varepsilon_i + \varepsilon' - \varepsilon_i)\frac{d\sigma}{d\varepsilon d\varepsilon'}d\varepsilon d\varepsilon' + 2\sum_{f\neq i}\int_0^{E_0+\varepsilon_i}(\varepsilon_f - \varepsilon_i + \varepsilon - \varepsilon_i)\frac{d\sigma_{fi}}{d\varepsilon}d\varepsilon, \end{aligned} \tag{3}$$

where the individual terms represent the contribution to the stopping cross section from single excitation, single ionisation, double excitation, double ionisation, and ionisation with excitation, respectively.

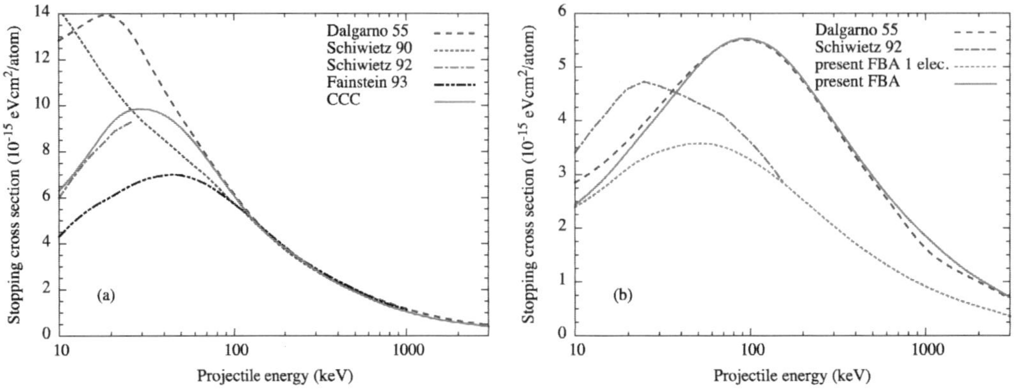

Figure 1. The stopping cross section for a beam of protons (a) and hydrogen atoms (b) passing through hydrogen. The present results for protons and hydrogen atoms are obtained using the CCC method and FBA, respectively. The theoretical calculations by Dalgarno and Griffing [2], Schiwietz [3], Schiwietz and Grande [5], and Fainstein *et al.* [6] are also shown.

3. Results

In Fig. 1a we show our present results of the stopping cross section for a beam consisting entirely of protons. As mentioned above these calculations were performed using the two-centre CCC method, which explicitly includes rearrangement processes. We included a basis of 20 *s*-states, 19 *p*-states, and 18 *d*-states on both projectile and target centres. We obtain good agreement with other theories above 100 keV. We also see reasonably good agreement with the AO+ results of Schiwietz and Grande [5]. In Fig. 1b we show our FBA results of the stopping cross section for a beam consisting entirely of hydrogen atoms. These calculations take into account single and double excitation and single and double ionisation as well as simultaneous excitation with ionisation. For excitation processes transitions up to the $n = 3$ shell were included. As expected our Born calculations are in good agreement with the similar calculations of Dalgarno and Griffing [2]. Schiwietz [3] used a static AO method to calculate the stopping cross section including only single-electron transitions. These calculations show a significant increase over the FBA calculation involving one-electron processes below 100 keV.

In order to calculate the total stopping cross section the present results shown in Fig. 1 need to be combined appropriately as detailed earlier. To this end we need to calculate the charge-state fractions. This in turn requires the knowledge of the electron-capture cross section in proton-H collisions and the electron-loss cross section in H-H collisions. The electron-capture cross section is the sum of transitions to all negative-energy projectile states, while the electron-loss cross section is the sum of the cross sections for single and double ionisation, and ionisation with excitation. The calculated charge-state fractions are presented in Fig. 2a. The experimental results of Allison [4] obtained using a H_2 target are also shown. In Fig. 2b we show our results for the total stopping cross section, which is the combination of two-centre CCC calculations and present FBA calculations. Despite the first-order approximation being used in the hydrogen-hydrogen channel, overall we obtain good agreement with experiment above 100 keV. However, at intermediate energies our results slightly overestimate experiment and at low energies they significantly underestimate experiment. Work is underway to eliminate approximations used in the hydrogen-hydrogen channel.

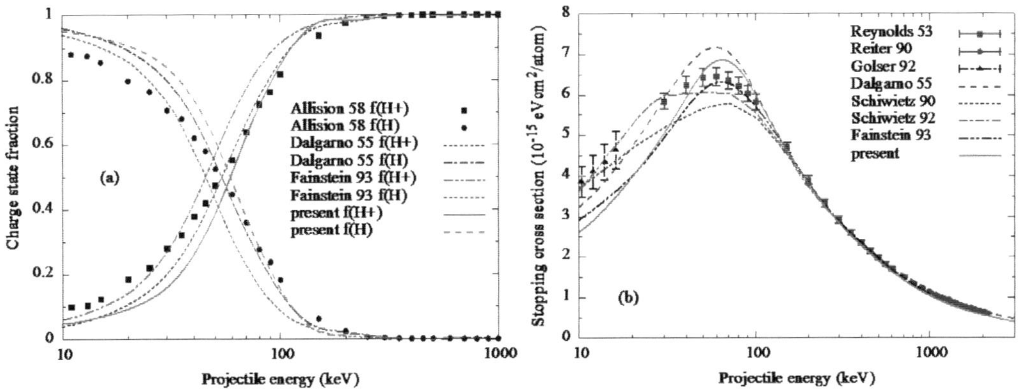

Figure 2. Charge-state fractions (a) and the total stopping cross section (b) for the proton-hydrogen collision system. The present results are obtained using a combination of the CCC method and FBA. Experimental measurements of the charge-state fractions have been performed by Allison [4]. The total stopping cross section have been measured by Reynolds *et al.* [11], Reiter *et al.* [12], and Golser and Semrad [13]. The theoretical calculations by Dalgarno and Griffing [2], Schiwietz [3], Schiwietz and Grande [5] and Fainstein *et al.* [6] are also shown.

4. Conclusion

In conclusion, we have presented a hybrid approach to calculating proton stopping in hydrogen. The stopping cross section is calculated using the two-centre CCC method for positive charge-state fraction and the Born approximation for the neutral charge-state fraction. Overall good agreement with experiment is obtained, however more theoretical work is required for complete solution of the problem at low energies.

5. Acknowledgments

This work was supported by Australian Research Council, Pawsey Supercomputer Centre, and National Computing Infrastructure. A.S.K. acknowledges a partial support from the U.S. National Science Foundation under Award No. PHY-1415656.

6. References

[1] McNamara A L *et al* 2012 *Int. J. Rad. Biol.* **88** 164-70
[2] Dalgarno A and Griffing G W 1955 *Proc. R. Soc. London A* **232** 423-34
[3] Schiwietz G 1990 *Phys. Rev. A* **42** 296-306
[4] Allison S K 1958 *Rev. Mod. Phys.* **30** 1137-68
[5] Schiwietz G and Grande P 1992 *Nucl. Instr. Meth. Phys. Res. B* **69** 10-7
[6] Fainstein P D *et al* 1993 *Phys. Rev. A* **47** 3055-61
[7] Bailey J J *et al* 2015 *Phys. Rev. A* **92** 022707
[8] Bailey J J *et al* 2015 *Phys. Rev. A* **92** 052711
[9] Avazbaev S K *et al* 2016 *Phys. Rev. A* **93** 022710
[10] Abdurakhmanov I B *et al* 2016 *J. Phys. B* **49** 03LT01
[11] Reynolds H K *et al* 1953 *Phys. Rev.* **92** 742-8
[12] Reiter G *et al* 1990 *Nucl. Instr. Meth. Phys Res. B* **44** 399-411
[13] Golser R and Semrad D 1992 *Nucl. Instr. Meth. Phys. Res. B* **69** 18-21

Synchrotron activation radiotherapy: Effects of dose-rate and energy spectra to tantalum oxide nanoparticles selective tumour cell radiosensitization enhancement

E Engels[1,2], M Lerch[1,2], M Tehei[1-3], K Konstantinov[2,4], S Guatelli[1], A Rosenfeld[1,2], S Corde[1,2,5]
[1]Centre for Medical Radiation Physics, University of Wollongong (UOW)
[2]Illawarra Health and Medical Research Institute (IHMRI), University of Wollongong
[3]Centre for Medical and Molecular Bioscience (CMMB), University of Wollongong
[4]Institute for Superconducting and Electronic Materials (ISEM), University of Wollongong
[5]Radiation Oncology Department, Prince of Wales Hospital, Randwick, NSW, Australia

E-mail: scorde@uow.edu.au

Abstract. Synchrotron radiation is unique in its ability to deliver dose at high dose rates using kiloelectronvolt photons. We are investigating the use of Tantalum pentoxide (Ta_2O_5) nano-structured particles (NSPs) that are to date unexplored in synchrotron radiation fields as they have high atomic number (Z=73) are biocompatible and are therefore potential radio sensitizers. We exposed cell culture flasks containing 9L gliosarcoma tumour cells or Madin-Darby Canine Kidney (MDCK) non-tumour cells to the NSPs and treated the cells using a broad synchrotron beam (140 keV median energy; average dose rate of 50 Gy/s) at the Australian Synchrotron. We compare the results with those from similar cells treated using a conventional 150 kV_p orthovoltage field (dose rate of 0.0127 Gy/s). The results reveal that the high dose-rate synchrotron irradiation is more effective at killing the 9L cells relative to the MDCK cells than the orthovoltage irradiation. On the other hand, the NSPs are more effective at radiosensitizing the 9L cells compared to the MDCK cells in the orthovoltage radiation field, which is due to the NSP energy dependence in the kilovoltage energy range. Both the dose rate and energy spectrum need to be considered in future studies with synchrotron activation radiotherapy (SART).

1. Introduction

High-Z nanoparticles (NPs) have been shown to improve the effectiveness of dose conformity to tumour tissue, particularly with macroscopic (broad beam) kilovoltage x-rays [1, 2]. The advantages of NPs, besides the primary benefit of enhancing local tumour radio-sensitivity, often include biocompatibility, permeability to cell membranes due to their nano-scale dimensions, and engineered abilities to specifically target certain tumours when coated.

Tantalum pentoxide (Ta_2O_5) nano-structured ceramic particles (TaNSPs) were first reported as radio-sensitizers by Brown *et al.* 2014 [2]. Nano-structured ceramic particles (NSPs) have different characteristics and furthermore, aggregation properties than inert metal NPs. TaNSPs have a high atomic number (Z=73), are biocompatible, and its oxide component may provide a useful target for drug attachment.

Polychromatic synchrotron radiation has significantly high dose rates, of order of tens to hundreds of Gy per second. The effect of these beams on tumour tissue are not yet well documented, however it

Journal of Physics: Conference Series **777** (2016) 012011 doi:10.1088/1742-6596/777/1/012011

is known that the inclusion of a high-Z targeting agents, such as platinum, gold, and iodine, can improve the treatment efficiency of synchrotron treatments, in their lower dose-rate monochromatic mode [3].

We aim to provide preliminary evidence of increased tumour response to high dose rates when compared to normal cells and characterise the ability of high-Z particles to selectively enhance synchrotron therapies, determining their efficacy toward Synchrotron Activation Radiotherapy (SART).

2. Methods

Tantalum pentoxide NSPs were synthesized according to the method detailed by Brown *et al.* [2] with sonication in an ultrasonic water bath, whilst suspended in PBS (phosphate buffered saline), for 2 hrs and mixed at 30 min intervals to promote the formation of smaller aggregates.

The NSPs were added to T12.5cm^2 cell culture flasks containing monolayers of 9L gliosarcoma cancer cells (derived from rodent brain cells) or MDCK (Madin-Darby Canine Kidney) cells 24 hours before 90-100% confluence. Each cell culture was grown with complete Dulbecco's modified eagle medium (c-DMEM) (DMEM supplemented with 10% fetal bovine serum (FBS) and 1% penicillin and streptomycin) and incubated at 37 °C and 5% (v/v) CO_2. NSP concentrations ranging from 50-500 µg/mL were added into flasks containing 5 mL DMEM.

Following 24 hrs of incubation with the NSPs, the cells were subcultured into 100 mm diameter petri dishes with 10 mL c-DMEM to evaluate NSP toxicity. Following 15 cell doubling times the cells were washed with 5 mL PBS (with calcium and magnesium ions) and fixed and stained with 5 mL of a (25% crystal violet and 75% ethanol) solution. Colonies consisting of no less than 50 cells were counted as surviving. The plating efficiency was evaluated as the ratio between the number of colonies present to the original cell seeding number. In this way, the impact of the irradiation or NSP toxicity was evaluated by comparing the plating efficiency of these treatments directly to the control plating efficiency and expressing this ratio as a normalized survival fraction.

For all experiments the flasks were irradiated when embedded in a 20 cm x 20 cm x 12 cm solid water phantom (RMI-457) with the cells at 2.5 cm depth. The irradiation field size was large enough to cover the whole flask in the case of the orthovoltage irradiations. For the synchrotron irradiations a step and shoot technique was utilized to create the irradiation field from the 1 mm x 12 mm intrinsic radiation field. The uniformity of the irradiation field was checked using EBT-3 film. The total dose was confirmed via a PTW Pinpoint™ ion chamber (model 31014).

Flasks with or without 50 µg/mL of NSPs were irradiated in a vertical orientation at hutch 1B on the Imaging and Medical Beamline (IMBL) at the Australian Synchrotron, with a single dose (2 Gy), at 50 Gy/sec, using a pink beam (4.65 mm Cu ex vacuo filtration). These results were compared with 2 Gy low dose-rate (LDR) 150 kVp orthovoltage X-ray beam from a Nucletron Oldelft Therapax DXT 300 Series 3 Orthovoltage unit at 0.76 Gy/min.

The differences in energy spectra of these beams is shown in Figure 1.

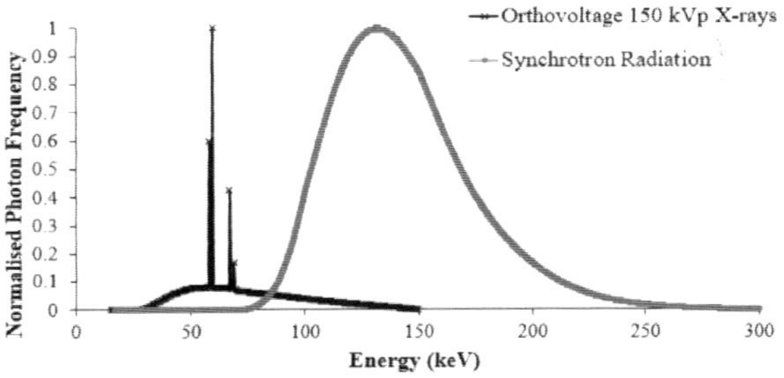

Figure 1. Photon beam energy spectra of 150 kVp orthovoltage X-rays (66 keV effective energy) and synchrotron pink beam (137 keV weighted average energy).

3. Results

Preliminary observations of TaNSPs with a light microscope, show they are distributed quite homogeneously amongst MDCK flask, except for some aggregation outside the cells over time (Figure 2). However, the TaNSPs tend to form clusters around the cell nucleus in the 9L gliosarcoma cells (Figure 3) due to the entrance of the TaNSPs into the cell and subsequent aggregation.

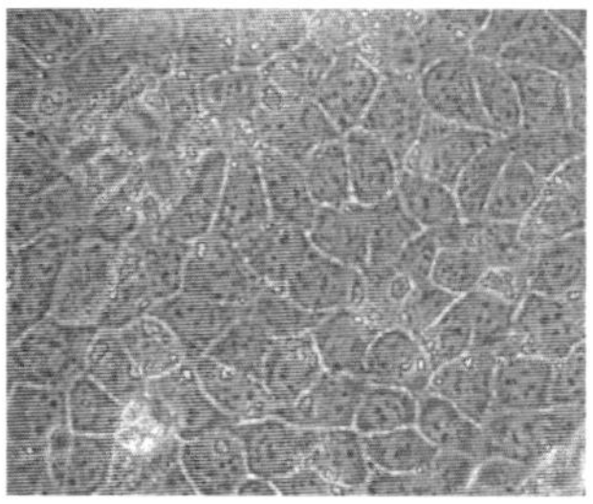

Figure 2. Ta_2O_5 NSPs (black-brown) in MDCK.

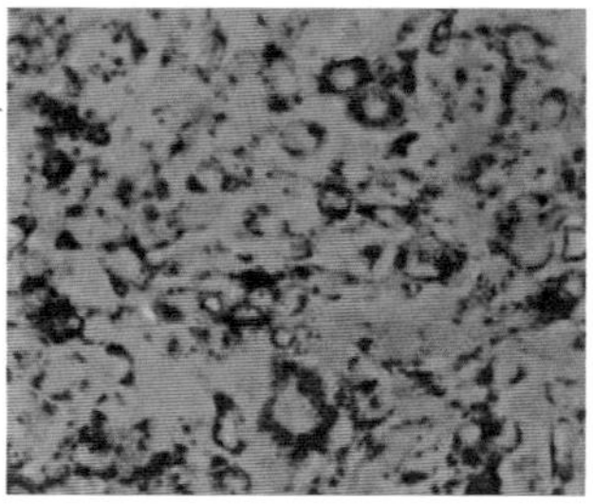

Figure 3. Ta_2O_5 NSPs (black) in 9L cells.

Toxicity of the Ta_2O_5 NSPs produce population survivals of 80-100%, in both cell lines without discrimination toward either cell line, as shown in Figure 4 for a range of NP concentrations.

In the case of 9L cells, the results are very consistent with previously published data from a different production batch and indicates the excellent reproducibility of the production technique.

This indicates that with TaNSPs, no toxic effect is acting on either cell line, and the cell survival following irradiation is purely dependent on the efficacy of the treatment itself.

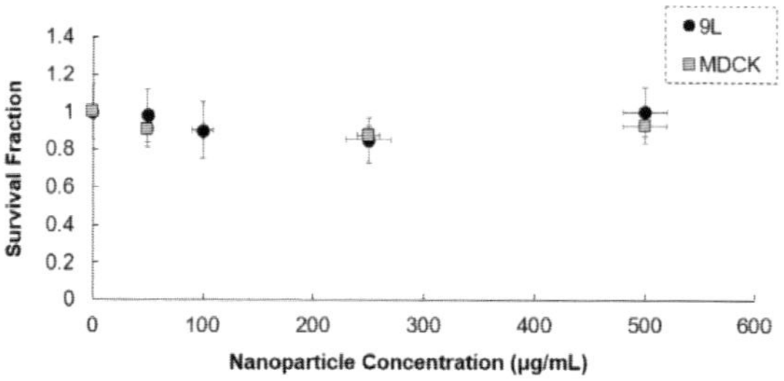

Figure 4. Intrinsic toxicity of Ta_2O_5 NSPs for 9L and MDCK cell lines.

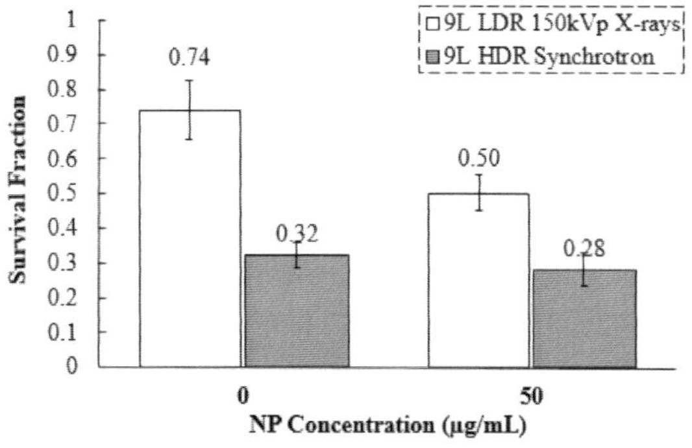

Figure 5. 9L cell survival following orthovoltage or synchrotron broad beam irradiation, with and without NSPs.

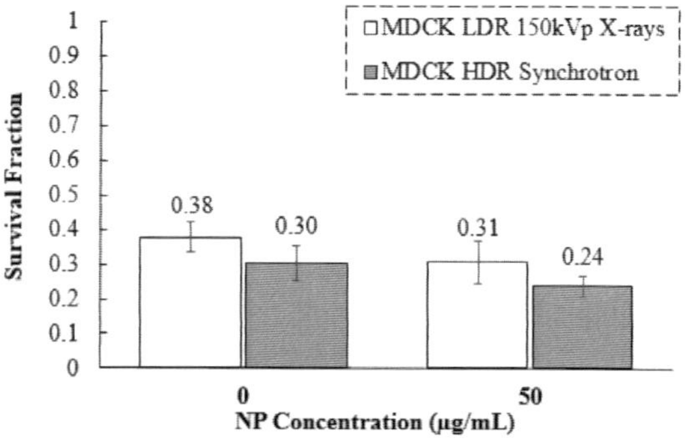

Figure 6. MDCK cell survival following orthovoltage or synchrotron broad beam irradiation, with and without NSPs.

Figures 5 and 6 highlight the effects of the different dose-rates on the cancer and normal cell lines respectively. Conventional 150 kVp orthovoltage treatment at 2 Gy LDR (low dose-rate) is compared with a synchrotron HDR (high dose-rate) broad beam.

Radioresistant 9L gliosarcoma cells appear to be more affected by the high dose rate than the MDCK cells. The orthovoltage treatment in comparison produces much greater 9L survival at 2 Gy than the synchrotron radiation. Radiosensitive MDCK in contrast, shows that there is insignificant difference between the orthovoltage treatment and HDR synchrotron treatment.

The addition of NSPs however, do not appear to significantly enhance the treatment of the 9L cells using synchrotron radiation. In contrast, the orthovoltage treatment does see a noticeable dose enhancement expressed through an increase in cell death in the tumour cells. In the normal cells however, there is no therapy enhancement in either LDR or HDR treatment. We believe this is due to the difference in the nanoparticle uptake by the tumour line (more nanoparticles closer to the nucleus).

These results highlight the importance of considering the kV beam energy used for SART. NP dose enhancement is indeed highly energy dependent in the kilovoltage range.

Figure 7 shows that the 150 kVp X-rays was precisely chosen to more directly target the photon absorption of the TaNPs relative to water and therefore optimize cell sensitization enhancement. The broad beam synchrotron radiation spectra would need further optimization and be lowered in energy to fully take advantage of dose enhancement by photoelectric effect on TaNSPs.

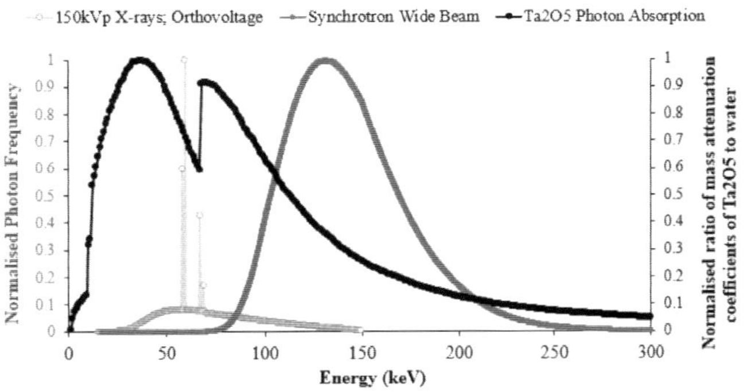

Figure 7. Comparison of beam energy spectra (grey) and ratio of total mass energy absorption coefficients of Ta_2O_5 to water (black) [4].

4. Conclusion

The inclusion of tantalum (V) oxide NSPs in kV beams is shown to improve tumour selectivity of the therapy. We believe the improvement is related to the preferential uptake of the NSPs by the 9L gliosarcoma cells compared to the MDCK kidney cells, which seems to naturally occur. The selective uptake leads to a higher concentration of NSPs in the 9L cells compared to the MDCK cells. The higher concentration also leads to selective dose enhancement in 9L cells due to the high-Z nature of the NSPs acting to significantly increase the interaction cross section of the cells at low photon energies (40-100 keV). Furthermore, the high dose rate of the synchrotron broad beam tends to produce greater effect on 9L radioresistant tumorous cells. However, NSPs are less effective in the synchrotron broad field for either cell line in this case, due to the beam energy spectrum requiring further optimization to maximize the radiosensitization of the TaNSPs.

5. Acknowledgments

The authors would like to acknowledge the members of staff at the Australian Synchrotron, and Prince of Wales Hospital for their time and resources; The Institute for Superconducting and Electronic Materials for the facilities to produce the nanoparticle; and the financial support of the Australian NH&MRC (APP1084994) and the Australian Synchrotron (AS152/IM/9404).

6. References

[1] Mesbahi A 2010 *Rep. Prac. Oncol. Radiother.* **15** 176-80
[2] Brown R *et al* 2014 *Part. Part. Syst. Charact.* **31** 500-5
[3] Bräuer-Krisch E *et al* 2015 *Phys. Medica* **49** 568-83.
[4] Hubbell J, Seltzer S, 2004, *Standard Reference Data Program of the National Institute of Standards and Technology (NIST)* (http://www.nist.gov/pml/data/xraycoef)

Advances in biological dosimetry

A Ivashkevich[1,2], T Ohnesorg[3], C E Sparbier[1], H Elsaleh[1,4]
[1]Radiation Oncology, Canberra Hospital, Garran, ACT, 2605, Australia
[2]Australian National University, Canberra ACT 2600, Australia
[3]Molecular Development, Murdoch Children's Research Institute, Melbourne, VIC, 3052, Australia
[4]College of Medicine Biology and Environment, Australian National University, Canberra ACT 2600, Australia

E-mail: alesia.ivashkevich@act.gov.au

Abstract. Rapid retrospective biodosimetry methods are essential for the fast triage of persons occupationally or accidentally exposed to ionizing radiation. Identification and detection of a radiation specific molecular 'footprint' should provide a sensitive and reliable measurement of radiation exposure. Here we discuss conventional (cytogenetic) methods of detection and assessment of radiation exposure in comparison to emerging approaches such as gene expression signatures and DNA damage markers. Furthermore, we provide an overview of technical and logistic details such as type of sample required, time for sample preparation and analysis, ease of use and potential for a high throughput analysis.

1. Introduction

As one of the most recent nuclear accidents, the Fukushima Daiichi nuclear disaster in 2011, clearly demonstrated, efficient contingency plans are required to assist with population triage in case of large scale radiation exposure. Biological dosimetry complements physical measurements and in some circumstances might be the only available approach to provide a reliable dose estimate. It is performed in combination with clinical assessment during the early response phase of a radiation emergency in order to retrospectively assess the imparted radiation dose and assist to determine the potential biological effects [1]. Biodosimetry methods are of great assistance for the evaluation of biological effects of the imparted radiation dose, although some caution is required for interpretation of the results. Increased individual susceptibility to radiation effects might result in a dose overestimation. Despite the well demonstrated usefulness of cytogenetic biodosimetry, the practical applications are limited by the time required to run the assay. Therefore, extensive developments and validation of novel assays were undertaken during the recent two decades to improve biodosimetry.

2. Current approaches, limitations and new developments in biodosimetry

2.1. Biomarkers in biodosimetry

The definition of a biomarker is 'any measurement reflecting an interaction between a biological system and an environmental agent, which may be chemical, physical or biological' [2]. Radiation biomarkers can be used for: 1) estimation or validation of received doses, 2) investigation of individual susceptibility, 3) early detection of radiation-induced health effects.

Conventional cytogenetic biodosimetry assays use chromosomal aberrations as indication of exposure, and scoring of dicentrics is considered the 'gold standard' of biodosimetry. Measured cytogenetic changes are the end-product of processing and cellular responses to inflicted damage. In contrast, emerging approaches, such as detection of DNA damage and gene expression changes, are indicative of initial cellular processes occurring shortly after exposure (Figure 1).

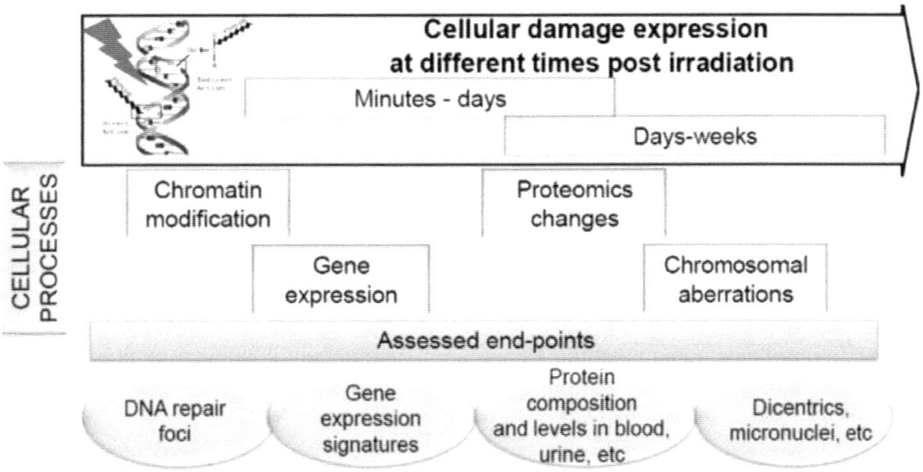

Figure 1. Overview of novel and existing biodosimetry methods.

Depending on the detected event in the sequence of DNA damage elicited molecular responses, current markers of radiation exposure can be broadly classified into:

- biomarkers detecting direct damage to DNA
- epigenetic modifications
- changes to gene expression (transcriptional) and altered protein composition and levels (translational changes)
- induced somatic and germline mutations/variants
- chromosomal changes as detected by cytogenetic biomarkers

2.2. *Ideal biomarker*

For a biomarker to be utilized, the assay accuracy of the dose estimate is of paramount importance to aid a reliable retrospective estimation of radiation doses in exposed individuals. Apart from being rapid, specific, sensitive, reproducible, an ideal bioassay should be able to distinguish the nature of exposure, and provide a reliable dose estimate irrespective of the exposure time. Biodosimetry assays also have to be high throughput and cost-effective.

It is also required to accurately measure and potentially differentiate between low and high dose, acute or prolonged, and external or internal exposure (Figure 2).

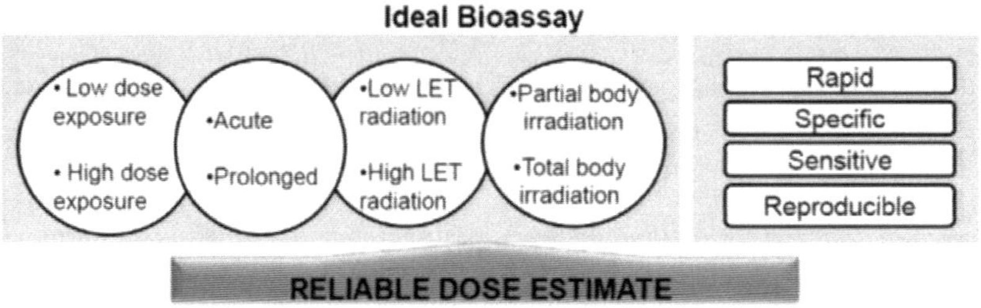

Figure 2. Essential features of an 'ideal' assay for optimal biodosimetry.

With a wide choice of methods, appropriate assays can be utilized depending on the situation. E.g. in emergency settings, for clinical management of accidental exposure, the expected dose assessment has to be accurate enough to predict clinical symptoms in order to appropriately treat the post accidental syndrome. Biological methods for detection of radiation found further applications in clinical settings for assessment of tumour and normal tissue sensitivity and response of tumours to chemoradiation [3-6].

2.3. Sample access and processing time
Biological material accessible for biodosimetry includes blood, urine, buccal cells, skin, hair follicles, etc. Peripheral blood lymphocytes (PBL) represent the most suitable sample as cells travel around the entire body; therefore, measurements performed in PBL enable the assessment of partial body exposure. About 2% of PBL are located in blood, continuously exchanging with those located in tissues and organs [6]. PBL can remain in circulation for several years allowing measurements for an extended time post exposure. PBL represent a suitable biological sample for all bioassays. However conventional cytogenetics takes longer than many other bioassays as the protocol requires mitogen (phytohemaglutinin) stimulation to activate proliferation of normally G0 lymphocytes. Cells have to undergo mitosis in order to exhibit damage detectable as translocations/micronuclei. Analysis of DNA damage repair foci (γH2AX) or gene expression changes is usually performed in PBL, too. These assays require no stimulation and the dose assessment can be performed within 12 hrs post exposure.

2.4. Dose response
Accurate dose quantification can only be performed using calibration dose response curves. Radiation dose dependent reproducible changes in marker expression, detectable with the selected assay are a prerequisite for successful biodosimetry methodology. Reduced sensitivity towards lower range doses, saturation of the response, and the detection threshold are potential limitations to consider. Saturation of the measured response limits method's ability to detect high doses and affect the time course of the response decay.

Dose response relationship for the yield of dicentrics produced by photon sources is fitted with a linear-quadratic model, which is consistent with single- and two-track hit models of aberration formation, $y = \alpha D + \beta D^2$ [7]. If exposure is spread over a certain period of time, damage repair will occur, and the quadratic component of the dose response relationship will gradually disappear. In chromosomal aberrations assay, the shape of the dose response curve depends strongly on Relative Biological Efficiency (RBE). Fission neutrons induced a significantly higher yield of dicentrics than that caused by low-Linear Energy Transfer (LET) sources. The linear component of the model, corresponding to damage caused by single-tracks, is predominant with fission neutrons. γH2AX foci show a linear dose response across a broad dose range [8, 9] and the assay is sensitive to doses as low as a few milligrays. In gene expression based biodosimetry, a linear dose-response relationship has been obtained for some genes over a dose range of 1–3 Gy [10], 0.5–4 Gy [11] and down to doses as low as 2 cGy [12].

2.5. In vivo validation of dose response

Radiation dose-response relationships are frequently analysed *in vitro* and have to be validated *in vivo*. Patients undergoing thyroid cancer treatment and total body irradiation during the treatment course for haematological malignancies [13] are main cohorts which were used to validate *in vitro* established assays. Although generally being in a good agreement with a physical dose, results of biological dosimetry tend to overestimate the actual physical dose [13]. Biodosimetry data relying on cancer patients have to be interpreted with caution, as it might not entirely reflect the situation in healthy cohorts due to the higher chance of patients carrying radiation susceptibility/resistance, and a potential exposure to previous chemo-radiation.

2.6. Discrimination between partial versus total body exposure

In case of chromosomal aberrations and γH2AX foci, homogeneous exposure yields a Poisson distribution of the aberrations and DNA repair foci, whereas inhomogeneous exposure results in contaminated Poisson distribution [14, 15]. Gene expression based biodosimetry approach lacks the capacity to distinguish this kind of difference in exposure.

2.7. High throughput approaches

The dicentric assay requires two days from sampling to results and it has limited sample processing capacity for mass screening following a nuclear incident. Standard biological dosimetry procedure (500 manually scored metaphases) is suitable for a few dose estimations and it cannot be performed in a timely manner in the case of a large-scale accident. A recent estimation, which included the 15 European Union countries where biological dosimetry is established, gave a total capacity for dosimetric triage of about 1500 cases per week [16]. Automated detection of dicentrics was introduced to enable a high throughput analysis. It has greatly reduced the time needed for dose estimation in whole- and partial-body accidental exposures without compromising its accuracy [7, 17]. Image analysis automation included development of a specialised software and algorithms for a high throughput counting of DNA repair foci [4].

2.8. Optimisation of gene expression biodosimetry for high-throughput applications

Detection of changes in gene expression has shown to be a promising biodosimetry approach [10, 18]. As a differential induction of alternative transcription in some genes (*MDM2, CDKN1a*) by irradiation is well documented [19, 20], one of the most recent studies suggested utilizing parts of the genes not induced by irradiation as intragenic control in order to decrease intra-individual variability in gene expression analysis [21].

Combination of automated biodosimetry assays is incorporated in so-called RABiT-infrastructure created to support ultra high throughput biodosimetry [22, 23].

3. Conclusions and Outlook

As each assay possesses certain limitations (scoring dicentrics is difficult to apply for triage of a large scale accident, the γH2AX assay is useful only within a short time window post exposure) and unique advantages (both γH2AX and gene expression analyses are sensitive and rapid bioassays), a combination of methods depending on the specific situation of a radiation accident will most likely allow the most accurate retrospective dose evaluation.

Despite recent advances and spurring developments of new assays, further improvement and validation will result in better reliability and accuracy of biodosimetry. Establishment of global infrastructure integrating new developments and access to required resources will assure better preparedness at all levels for every kind of nuclear disasters.

4. References

[1] International Atomic Energy Agency 1998 Diagnosis and treatment of radiation injuries *Safety Reports* **2** (Vienna, Austria).
[2] World Health Organisation 1993 Biomarkers and risk assessment: concepts and principles *EHC 155* (Geneva, Switzerland).
[3] Minicucci E M *et al* 2014 *World J. Clin. Oncol.* **5** 93-102
[4] Ivashkevich A N *et al* 2011 *Mutat. Res.* **711** 49-60
[5] Ivashkevich A *et al* 2012 *Cancer Lett.* **327** 123-33
[6] International Atomic Energy Agency 2011 Cytogenetic Dosimetry: Applications in Preparedness for and Response to Radiation Emergencies (Vienna, Austria)
[7] Prasanna P G S *et al* 2002 AFRRI's Gamma-Ray, X-Ray, and Fission-Neutron Calibration Curves for the Lymphocyte Dicentric Assay: Application of a Metaphase Finder System. *Armed Forces Radiobiology Research Institute* (Bethesda, Maryland)
[8] Beels L *et al* 2010 *Int. J. Radiat. Biol.* **86** 760-8
[9] Moroni M *et al* 2013 *Int. J. Mol. Sci.* **14** 14119-35
[10] Grace M B *et al* 2002 *Int. J. Radiat. Biol.* **78** 1011-21
[11] Kang C M *et al* 2003 *Radiat. Res.* **159** 312-9
[12] Amundson S A, Fornace A J, Jr. 2003 *Health Phys.* **85** 36-42
[13] Dossou J *et al* 2000 *Int. J. Radiat. Oncol. Biol. Phys.* **46** 123-9
[14] Hilali A *et al* 1991 *Radiat. Res.* **128** 108-11
[15] Redon C E *et al* 2010 *PLoS One* **5** e15544
[16] Wojcik A *et al* 2010 *Radiat. Prot. Dosimetry* **138** 397-401
[17] Vaurijoux A *et al* 2012 *Radiat. Res.* **178** 357-64
[18] Kabacik S *et al* 2011 *Int. J. Radiat. Biol.* **87** 115-29
[19] Sprung C N *et al* 2011 *PLoS One* **6** e25758
[20] Forrester H B *et al* 2012 *PLoS One* **7** e53358
[21] Forrester H B *et al* 2014 *Radiat. Res.* **181** 314-23
[22] Garty G *et al* 2011 *Int. J. Radiat. Biol.* **87** 754-65
[23] Turner H C *et al* 2014 *Radiat. Environ. Biophys.* **53** 265-72

Challenges in describing nuclear reactions outcomes at near-barrier energies

M Dasgupta, E C Simpson, Kalkal S, K J Cook, I P Carter, D J Hinde, D H Luong
Department of Nuclear Physics, Research School of Physics and Engineering, Australian National University, ACT 2601, Australia

E-mail: mahananda.dasgupta@anu.edu.au

Abstract. The properties of light nuclei such as ^{6}Li, ^{7}Li, ^{9}Be and ^{12}C, and their reaction outcomes are known to be strongly influenced by their underlying α-cluster structure. Reaction models do not yet exist to allow accurate predictions of outcomes following a collision of these nuclei with another nucleus. As a result, reaction models within GEANT, and nuclear fusion models do not accurately describe measured products or cross sections. Recent measurements at the Australian National University have shown new reaction modes that lead to breakup of ^{6}Li, ^{7}Li into lighter clusters, again presenting a further challenge to current models. The new observations and subsequent model developments will impact on accurate predictions of reaction outcomes of ^{12}C - a three α-cluster nucleus – that is used in heavy ion therapy.

1. Introduction

Energetic beams of ^{12}C, used in heavy ion therapy, undergo a range of reaction processes as they slow down in matter. This is schematically illustrated in Fig. 1. At high energies, knockout, fragmentation and spallation reactions occur, whilst at medium and low energies transfer, breakup and fusion reactions are dominant. No single model can describe the reaction outcomes over such a large energy range (~200 MeV/u to stopping), and thus different models are used depending on the energy of the ion, as shown in Fig. 1.

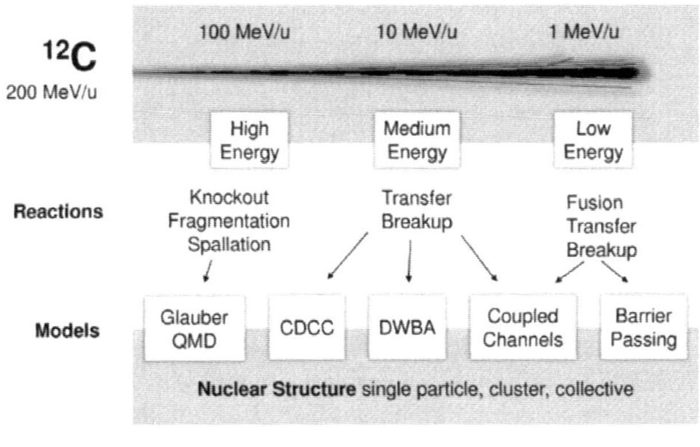

Figure 1. Schematic diagram showing the dominant nuclear reactions that occur as high energy carbon nuclei slow down in matter. Different models, each suited to a particular energy regime, are used. For example Glauber and quantum molecular dynamics (QMD) models are used at high energies, whilst continuum discretised coupled channels model (CDCC) and distorted wave Born approximation (DWBA) are used at medium energies. All models need to include the effects of nuclear structure.

The quantum structure of the nucleus plays an important role in reaction outcomes at all energies. Nuclear structure fundamentally arises from the strong interactions of the constituents of the nucleus. However, we cannot yet calculate all aspects of nuclear structure from first principles (i.e. starting from quarks). Thus structure models highlight specific aspects, e.g. of single particle nature or of collective nature where all constituents participate, leading to vibrations or rotations of the nucleus. The description of nuclear structure is an essential ingredient in all reaction models, as the reaction dynamics is strongly influenced by nuclear structure.

The nucleus ^{12}C is known to have an α-cluster structure [1], which fundamentally arises due to the large binding energy of the ^{4}He nucleus (α-particle). Interactions of energetic beams of ^{12}C with matter produces Li, Be and B fragments, which as well as carbon, contribute to the depth-dose distribution [2]. Production of α-particles is also observed but their yields at low energies are underestimated [2] by up to a factor of 10 in simulations. This underestimation has been attributed to effects of α-clusters in ^{12}C. Nuclear collisions near the Coulomb barrier are very sensitive to nuclear structure effects, and thus studies of reactions near the barrier can reveal the link between structure and reaction outcomes. This is particularly relevant for nuclei with α-clusters, as they can disintegrate via many mechanisms as shown by measurements in the last several years. This complexity means that some major reaction outcomes are not described by the current quantum models of nuclear reactions.

The experiments [3-6] and model developments [7] at the Australian National University (ANU) are addressing this challenge. The measurements [4-6] have concentrated on lithium – the next higher element to ^{4}He. The isotopes ^{6}Li and ^{7}Li have α-^{2}H and α-^{3}H ground state cluster structure. The reaction outcomes are influenced by their cluster structure, e.g. complete fusion cross sections are found to be reduced by ~30% compared with expectations [8, 9]. It was thought that this suppression arises due to breakup of lithium nuclei into their cluster constituents. However, experiments at the ANU, described below, show that breakup of lithium is more complex and can occur via many routes.

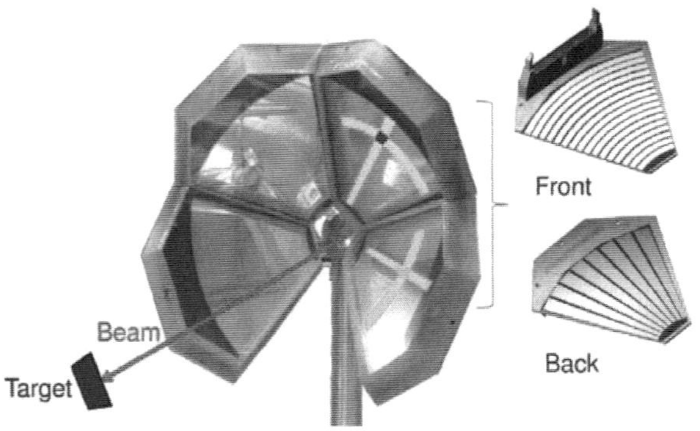

Figure 2. Four double-sided silicon strip detectors are used to detect coincident charged breakup fragments. The front of each detector is divided into sixteen arcs and the back is divided into eight sectors. Positon of each breakup fragment can be determined from the intersection of the arc and the sector, as illustrated in the figure. The figure shows a schematic sketch of the beam (red arrow), hitting the target foil, with two charged fragments being registered at backward angles (purple and yellow pixels) by the detector array.

2. Experimental Methods

The experiments were performed using beams of ^{6}Li and ^{7}Li from the 14UD electrostatic accelerator, at the Heavy Ion Accelerator Facility at the ANU. Beams were incident on thin isotopically enriched targets. The evolution of reaction mechanisms with decreasing target mass was studied by using targets ranging from ^{209}Bi via ^{208}Pb, ^{144}Sm, ^{64}Zn, ^{58}Ni, ^{27}Al, ^{12}C, to ^{2}H. Interactions of the $^{6,7}Li$ beams with the target nuclei at energies below the barrier result in coincident charged breakup fragments (e.g. ^{4}He, ^{3}H, ^{2}H, and ^{1}H). The coincident charged breakup fragments were detected using an array of four 400 μm thick double-sided silicon-strip detectors. The position information of the coincident fragments, together with their measured energies were used to reconstruct the kinematics [3-6]. Thus the energetics of the reactions (Q-values) could be determined, giving information on the state of the heavy (target-like) nucleus. Newly identified experimental variables that give information on the position of breakup have recently been determined [9], leading to an understanding of the reaction dynamics with in detail.

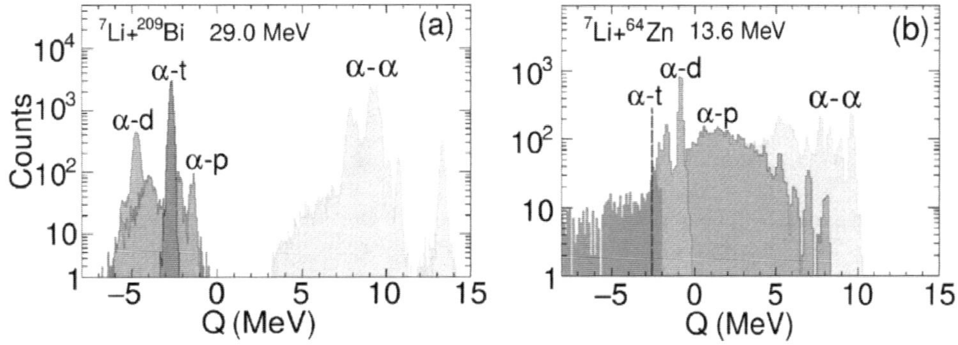

Figure 3. Breakup of ^{7}Li nucleus, which has an α-^{3}H cluster structure, proceeds via many modes. The various breakup modes are clearly seen in the Q-value plot shown for measurements with targets of (a) ^{209}Bi and (b) ^{64}Zn. Breakup into α and ^{3}H (triton, t), shown in blue, is observed for the heavy target, but not the lighter target – the vertical blue line in (b) indicates the expected Q-value. Breakup following transfer of nucleons is dominant. Proton transfer from the target to ^{7}Li leading to ^{8}Be, which breaks up into two alpha particles, is shown in green; one and two neutron transfer from ^{7}Li forms ^{6}Li and ^{5}Li, which then breakup into α-deutron (α-d) and α-proton (α-p) pairs respectively.

3. Results

The experiments show that for ^{6,7}Li induced reactions, in addition to direct breakup into α-d and α-t clusters respectively, breakup via other modes is very significant [4-6]. These modes arise following transfer of nucleons (protons or neutrons) between the colliding nuclei that can populate nuclei that are themselves weakly bound or unbound. These modes are present in interactions with nuclei ranging from bismuth through nickel down to deuteron; the breakup modes for bismuth and zinc are shown in Fig. 3. The figure shows that the direct breakup component (that can, in principle, be described theoretically) is negligible for interactions with Ni; essentially the direct breakup component is absent for collisions with lower mass target nuclei [6].

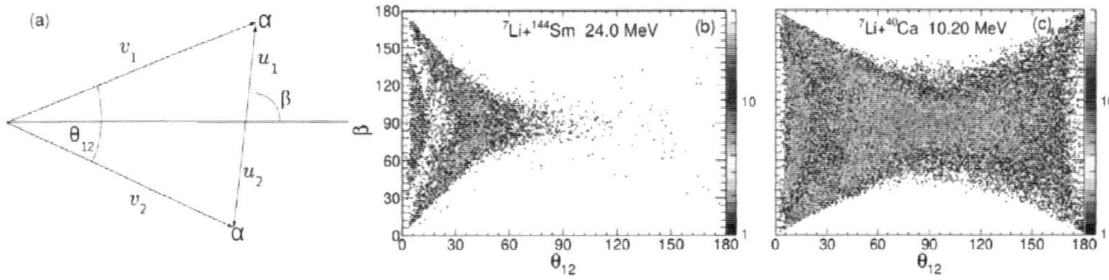

Figure 4. The angle between the two α-particles θ_{12}, following the disintegration of ^{8}Be → α + α, as a function of the angle β for the indicated targets. The relationship between angles θ_{12} and β is sketched in panel (a). The angle at which the α-particle emerges depends on the location of breakup and the atomic number of the target.

Helped by classical breakup simulations, experimental observables have been found that are sensitive to the breakup location [9], i.e. whether breakup occurs close to the target or asymptotically far from the target. The breakup location depends on the meanlife of the parent state; sensitivity to subzeptosecond lifetimes is found. An example is shown in Fig. 4, for breakup of 7Li following proton pickup from 144Sm and 40Ca; the nucleus 8Be is formed which subsequently breaks up into two α-particles. In the context of heavy ion therapy, the location of the breakup and the atomic number of the target affect the angular distribution of the breakup fragments.

The figure shows the angle between the α-particles θ_{12} against β which is the orientation of the relative velocity (of the two α-particles) with respect to the velocity of the center of mass of the fragments [sketched in Fig. 4 (a)].

4. Conclusions

The experiments demonstrate that breakup in reactions of ^{6,7}Li is largely triggered by transfer. The dominance of transfer triggered breakup increases as the target mass is lowered. These transfer triggered breakup mechanisms are not currently incorporated in models (nor in simulations) of nuclear reactions near the barrier. The experimental results necessitate new developments in modelling nuclear reactions. The experimental advances also pave the way to study the breakup mechanisms of other nuclei such as ^{12}C, ^{13}C and light radioactive nuclei such as ^{6}He, ^{8}Li, ^{11}Li, ^{11}Be etc. that are beginning to be available from rare isotope accelerator facilities world-wide.

5. Acknowledgments

Support for operations of the ANU Heavy Ion Accelerator Facility through NCRIS is acknowledged. This work was supported by ARC grants DP0879679, FL11010098, DP130101569, DE140100784.

6. References

[1] Freer M 2007 *Rep. Prog. Phys.* **70** 2149
[2] Nuclear physics for medicine, NuPECC report (2014) (www.nupecc.org/npmed/npmed2014.pdf, pg 39)
[3] Rafiei R *et al* 2010 *Phys. Rev.* C **81** 024601
[4] Luong D H *et al* 2011 *Phys. Lett.* **695** 105
[5] Luong D H *et al* 2013 *Phys. Rev.* C **88** 034609
[6] Kalkal S *et al* 2016 *Phys. Rev.* C (in press)
[7] Diaz-Torres A *et al* 2007 *Phys. Rev. Lett.* **98** 152701
[8] Dasgupta M *et al* 2002 *Phys. Rev.* C **66** 041602(R)
[9] Simpson E C *et al* 2016 *Phys. Rev.* C **93** 024605

Innovative detectors for quality assurance dosimetry in SBRT of stationary and movable targets

M K Newall[1], M Petasecca[1], M Duncan[1], I Fuduli[1], K Al shukaili[1], J T Booth[2], P Keall [2,3], S Corde[4,1], V Pereveratylo[5], M Lerch[1], A B Rosenfeld[1]
[1]Centre for Medical Radiation Physics, University of Wollongong, Australia
[2]North Shore Cancer Care Centre, Australia
[3]Radiation Physics Laboratory, University of Sydney, Australia
[4]Department of Radiation Oncology, Prince of Wales Hospital, Australia
[5]SPA-BIT, Kiev, Ukraine

E-mail: anatoly@uow.edu.au

Abstract: The high spatial and temporal resolution 2D monolithic silicon detector arrays M512 and DUO for quality assurance (QA) in real time motion adaptive radiotherapy (ART) have been developed. The DUO array possesses a spatial resolution of 0.2 mm and has demonstrated agreement within 5% with EBT3 film measurements of 6MV linac beam profiles for field sizes 1 × 1 cm^2 and SRS cone diameter 0.5 cm. Dynamic characterisation of the M512 for QA in real time ART evaluated the performance of M512 for small fields while the detector experiences periodic motion. It was demonstrated with M512 that MLC tracking with Calypso electromagnetic array compensates for the periodic motion and improves the agreement between static and dynamic beam profiles for field size 1 × 1 cm^2 from within 75% in the penumbra to within 11% agreement. The dynamic profile is returned to a similar distribution as the static case.

1. Introduction

Stereotactic Radiotherapy (SRT) describes a form of external beam radiotherapy (EBRT) treatment delivering a hypo-fractionated high dose regimen using highly conformal photon beams. Stereotactic Body Radiation Therapy (SBRT) is the application of this methodology to extracranial tumour sites with tumour diameters less than 5 cm [1]. Treatment sites may include: lung, liver and kidneys within the abdominal and thoracic cavities [2]. SBRT requires accurate delineation of the target structure and minimisation of treatment margins associated with physical, biological and physiological tumour processes and motions, to spare surrounding healthy tissue and nearby critical organs [3]. Accurate delineation is achieved through pre-treatment imaging protocols and improved over the course of treatment using a combination of image guidance and motion management strategies.

The treatment of small volume tumours in the vicinity of critical organs within the abdomen requires conformal small area radiation beams [4, 5]. Challenges in small field dosimetry are well known [6, 7] and require that an ideal dosimeter for small area radiation beams exhibits properties including: dose-rate, energy and directional independence, near water equivalence, minimal beam perturbation and small sensitive volume [8, 9].

The radiobiological advantages associated with adopting highly conformal, hypo-fractionated, high dose treatments, are compromised by the impact of tumour motion (both inter and intra-fraction). Lung tumours may move up to 20-30 mm (superior-inferior plane) during the respiratory cycle [10]. In general, to account for tumour motion, the treatment margins must be enlarged to encompass the full range of motion of the intended target, however, this sacrifices the conformal delivery of the treatment and increases dose delivery to surrounding healthy tissue.

Adaptive radiotherapy (ART) which minimises the need for large treatment margins based on the motion management strategies designed to minimise the effects of organ motion upon treatments include: gating [11], patient breath-hold and real-time tumour tracking with Multi-Leaf Collimator (MLC) adaptation [12].

The quality assurance (QA) in real time motion ART requires the dosimetry system possess both high spatial and high temporal resolution to accurately and simultaneously evaluate the delivered dose distribution and the consequences of the interplay effects attributable to the motion and tracking.

2. Materials and Methods

1.1. Detectors

The Centre for Medical Radiation Physics (CMRP), University of Wollongong, has developed novel pixelated silicon detectors, for use in the QA of modern EBRT. The MagicPlate 512 (M512) is a monolithic two-dimensional square array consisting of 512 silicon diodes with 2 mm pitch and 0.5 × 0.5 mm^2 sensitive area, figure 1(a). DUO is comprised of 505 silicon diodes arranged in intersecting orthogonal linear arrays with sensitive area 0.02 × 1 mm^2 and pitch 0.2 mm, figure 1(b).

(a)

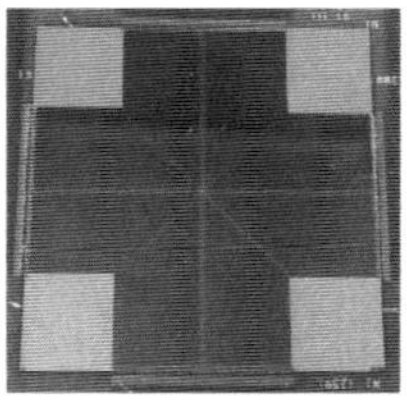

(b)

Figure 1. The silicon diode arrays mounted and wire bonded to printed circuit board carries. (a) MagicPlate 512. (b) DUO.

1.2. Characterization for small static fields

Beam profiles of a small area radiation field of size 1 × 1 cm^2 defined by the collimation jaws, with the MLC retracted and a SRS cone with diameter 0.5 cm, were measured. DUO was placed at a depth of 10 cm in a solid water phantom with 10 cm of backscattering material. The system was irradiated at 90 cm SSD using a 6 MV photon beam from a linac for 100 monitor units (MU) at 600 MU min^{-1}. The measurements were repeated under identical conditions with Gafchromic EBT3 film (ASHLAND, Wayne, NJ) for comparison.

1.3. Calypso Motion Tracking Array and MLC Tracking

MLC tracking utilising a Calypso (Varian Medical Systems) electromagnetic array and radiofrequency beacons implanted within the target, presents an attractive motion management solution [13]. The changing spatial coordinates of the target (or a surrogate) are provided to the MLC controller to alter the leaf positions in real time.

1.4. Characterization for dynamic target

Dynamic characterisation of the M512 for QA in real time ART evaluated the performance of M512 for small fields while the detector experiences periodic motion. M512 is placed upon a HexaMotion platform (ScandiDos, Uppsala, Sweden) on the treatment couch of a linac, figure 2 (a). The M512 is irradiated by square MLC defined fields of size 1 × 1 cm^2 at 1.5 cm water-equivalent depth and 100 cm SAD for 1000 MU at 600 MU/min, with the gantry at 0° rotation [14]. Three cases were investigated; static platform without MLC tracking, dynamic platform (lung motion) without MLC tracking and dynamic platform with MLC tracking (MLC compensates for motion of the target).

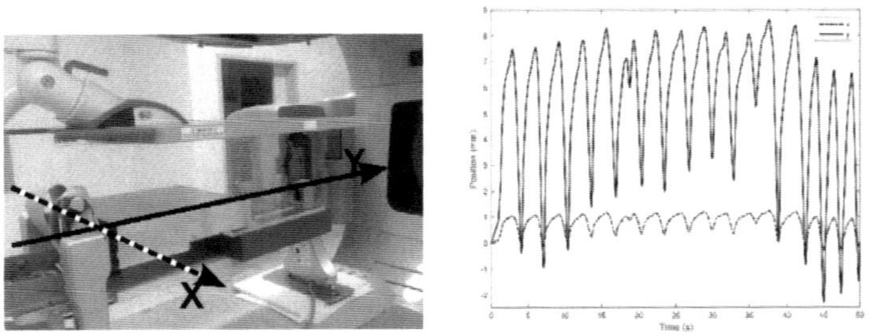

(a) (b)

Figure 2. (a) M512 detector positioned upon HexaMotion platform. (b) The temporal motion pattern provided to the platform. The platform mimics the lung motion during the irradiation across the two axis of motion provided *X* and *Y*.

3. Results

1.5. Characterization of small static fields

The EBT3 film response is normalised to the pixel values within a 1 mm window surrounding the central axis (CAX) of the profile. The DUO profiles are normalised to the response of the central detector pixel. The beam profiles acquired from both detectors are aligned such that the origin lies at the coordinate corresponding to 50% response for DUO and the EBT3 film.

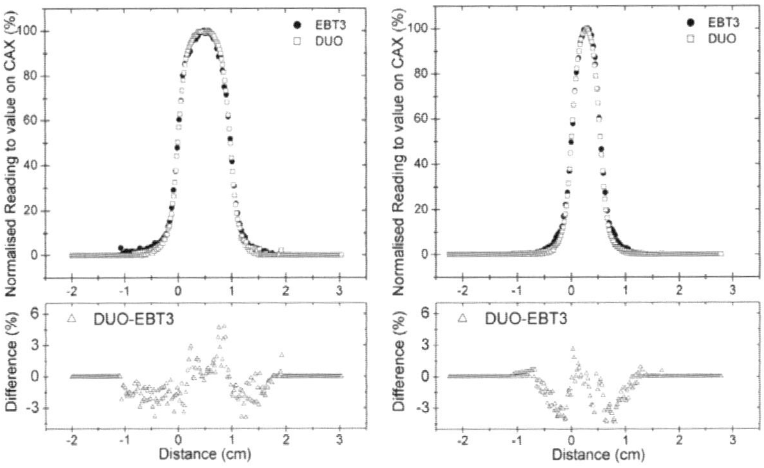

(a) (b)

Figure 3. Comparison of response of DUO detector and EBT3 film for small field sizes. (a) Field size 1 × 1 cm^2. Lower panel shows agreement between EBT3 and DUO. (b) SRS cone collimator 0.5 cm diameter. Lower panel shows agreement between EBT3 and DUO.

In Figure 3, DUO has demonstrated good agreement with EBT3 film measurements to within 5% in the penumbra of 6MV linac beam profiles for field sizes 1 × 1 cm^2 and SRS cone diameter 0.5 cm, figure 3(a), 3(b).

1.6. Characterization of dynamic target

The beam profiles of the M512 detector for the separate motion cases are compared, the line profiles are extracted along the identical column of channels within the array. The result shown in figure 4 illustrates the beam profiles along the *Y* direction of motion, figure 2 (a). M512 has been shown to agree with EBT3 film to within 0.4 mm for penumbral width measurements and 3% for FWHM measurements [14].

The profiles acquired with M512 for the motion cases along the *Y* axis is shown in figure 4. The dynamic platform without MLC tracking case demonstrates a shift in the delivered dose distribution. The motion introduces this displacement and smears the dose out in the penumbra leading to disagreement between the cases. The dynamic platform with MLC tracking case highlights the capability of the MLC tracking system to compensate for the periodic motion and return the profile to a similar distribution as the static platform case. The agreement within the penumbra of the profiles acquired with M512 for the no motion and motion cases is improved from within 75% without MLC tracking to within 11% with MLC tracking, figure 4.

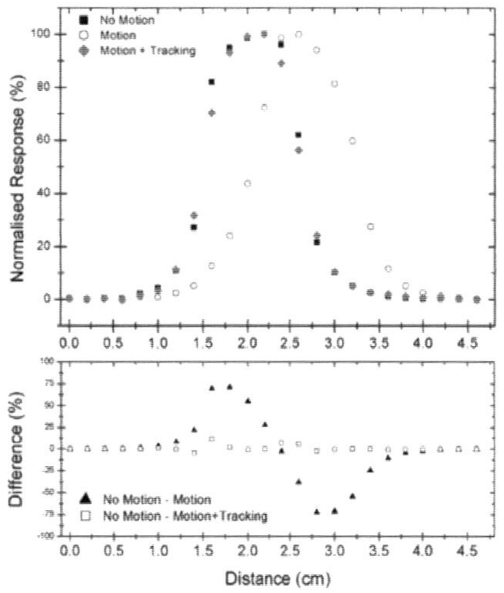

Figure 4. Comparison between responses from M512 for measurements of 1 × 1 cm^2 MLC defined field along the *Y* axis. M512 is irradiated in identical position for cases: i) static platform without MLC tracking – no motion case. ii) Dynamic platform without MLC tracking – with motion case. iii) Dynamic platform with MLC tracking – motion and tracking case. The lower panel shows the agreement between the no motion and motion case and the no motion and motion and tracking case, presenting the percentage differences between the opposing cases across the profiles.

4. Conclusion

M512 has been demonstrated as an attractive alternative to both radiochromic films and two-dimensional ionisation chamber arrays for the quality assurance of motion adaptive RT and SBRT [14]. The high spatial resolution of DUO presents valuable benefits for QA of real-time ART treatments. The increasing complexity of modern EBRT treatments and the further implementation of motion management strategies necessitates both high temporal resolution and real-time response in QA dosimetry systems to identify possible inaccuracy or failure in critical time-dependent effects.

5. References

[1] Hanna G G and Landau D 2015 *Clin. Oncol.* **27** 290-7
[2] Rubio C *et al* 2013 *Reports Pract. Oncol. Radiother.* **18** 387-96
[3] Kavanagh B D *et al* 2006 *Semin. Radiat. Oncol.* **16** 77-84
[4] Vial P *et al* 2008 *Med. Phys.* **35** 1267-77
[5] Hill R *et al* 2010 *Med. Phys.* **37** 4355-63
[6] Heydarian M *et al* 1996 *Phys. Med. Biol.* **41** 93-110
[7] Das I J *et al* 2008 *Med. Phys.* **35** 206-15
[8] Pappas E *et al* 2008 *Med. Phys.* **35** 4640-8
[9] Greer P B *et al* 2007 *Med. Phys.* **34** 4389-98
[10] Keall P J *et al* 2006 *Med. Phys.* **33** 3874
[11] Kubo H D and Hill B C 1996 *Phys. Med. Biol.* **41** 83-91
[12] Keall P J *et al* 2006 *Int. J. Radiat. Oncol. Biol. Phys.* **65** 1579-84
[13] Keall P J *et al* 2014 *Med. Phys.* **41** 020702
[14] Petasecca M *et al* 2015 *Med. Phys.* **42** 2992-3004

Monte Carlo characterisation of the Dose Magnifying Glass for proton therapy quality assurance

A H Merchant [2], S Guatelli[2], M Petesecca[2], M Jackson[2,3], A B Rozenfeld[2]
[2]Center for Medical Radiation Physics, University of Wollongong, Wollongong, Australia
[3]Radiation Oncology, Prince of Wales Hospital, Sydney, Australia

E-mail: ahm798@uowmail.edu.au

Abstract. A Geant4 Monte Carlo simulation study was carried out to characterise a novel silicon strip detector, the Dose Magnifying Glass (DMG), for use in proton therapy Quality Assurance. We investigated the possibility to use DMG to determine the energy of the incident proton beam. The advantages of DMG are quick response, easy operation and high spatial resolution. In this work we theoretically proved that DMG can be used for QA in the determination of the energy of the incident proton beam, for ocular and prostate cancer therapy. The study was performed by means of Monte Carlo simulations Experimental measurements are currently on their way to confirm the results of this simulation study.

1. Introduction

While proton radiotherapy has experienced a growth in the number of operational facilities and an increase in capabilities of beam delivery technology, there has been a substantial lack of development in commercial detectors for related Quality Assurance (QA). In particular, fast verification of the beam energy, range in water and position of the characteristic pristine or spread out Bragg Peak in both a pre-treatment and real time setting is necessary to prevent radiation accidents and deliver treatment to the patient safely, minimising the damage to healthy tissues and critical structures. Ideally detectors used in proton therapy QA should provide sub-millimetre spatial resolution determination of proton range over a wide energy range to resolve the steep energy gradient of the Bragg Peak and be able to provide this information at a sub-second temporal resolution.

Currently, the generation of available detectors implemented for proton therapy QA include ionization chambers, as recommended by the ICRU Report 78 and IAEA TRS-398 [1, 2], single diodes and films to derive the Bragg Peak position in water phantoms [3].

The Dose Magnifying Glass (DMG) is a silicon strip detector developed by the Centre of Medical Radiation Physics (CMRP) at the University of Wollongong. It is a miniature detector for use in proton therapy QA, with applications into X-ray stereotactic radiosurgery (SRT) and intensity modulated radiotherapy (IMRT) investigated previously demonstrating the DMG capability to provide sub-millimetre spatial resolution at a sub-second temporal resolution over a large (10^3) dynamic range [4]. Its advantages over existing commercial QA detectors include the simple device operation and high spatial resolution over a wide dynamic energy range, required to resolve the Bragg Peak accurately.

This paper presents the Monte Carlo characterisation of the DMG as device to determine the energy of the incident proton beam for ocular and prostate cancer proton therapy treatment. At this stage, the principle of operation is proved for mono-energetic proton beams only.

2. Dose Magnifying Glass

The DMG is a 128 channel p-silicon strip detector, with each strip consisting of a sensitive implanted n+ area of 20x2000 μm, separated by a pitch of 200 μm. This detector, in singular configuration, is mounted upon a thin Kapton carrier, as presented in figure 1, which provides a convenient connection of the 128 channels to a FPGA based multichannel electrometer, resulting in a total size of 4x28x0.25 mm. In serial configuration an additional detector, providing a further 128 channels, is positioned 4 mm behind the first, rotated 180° and translated such that the sensitive volumes remain aligned in both detectors. In this configuration the detectors are mounted upon a PCB board with the entire ensemble placed in a housing consisting of a 75x95x5 mm PMMA lower half and a RMI457 Solid Water top half with a wax recess for the detectors.

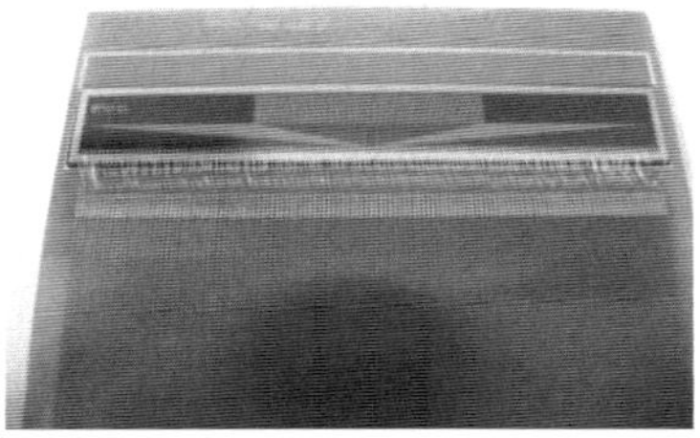

Figure 1. Singular configuration of the DMG mounted on a Kapton Carrier.

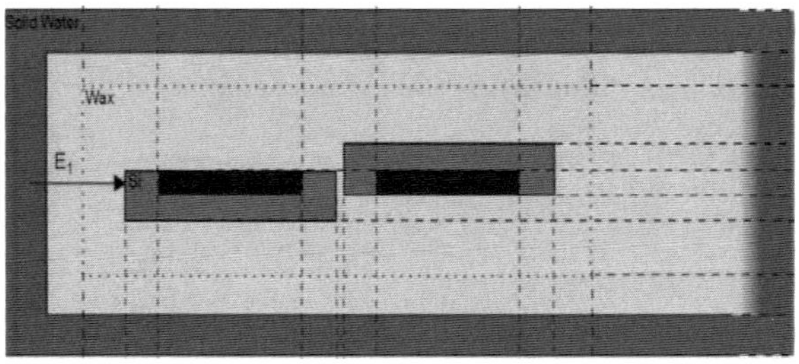

Figure 2. Serial configuration of the DMG mounted on a PCB board with wax recess and solid water housing shown.

3. DMG Principle of operation in pre-treatment QA to determine incident proton beam energy

When the device is positioned appropriately in the treatment target region in a water phantom, protons directly incident on the DMG produce a first Bragg Peak in the device. Other protons, instead, scatter in the phantom and in the detector packaging before entering in the DMG, producing a second Bragg Peak at a further depth. The distance between the two Bragg Peaks can be measured and uniquely related to the kinetic energy of the proton beam when incident on the device, using the CSDA range difference, shown in figure 3.

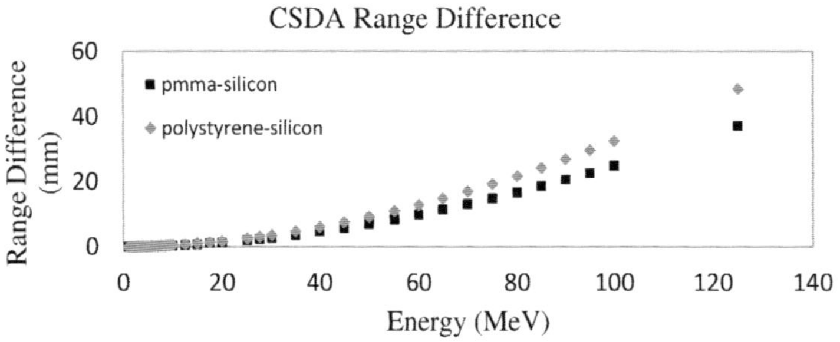

Figure 3. CSDA range difference of protons traversing silicon and PMMA/Polystyrene.

Once the energy E_{DMG} of the proton beam at the entrance of DMG is determined, the energy of the incident beam, $E_{incident}$ at the entrance of the phantom can be calculated by considering the associated CSDA range of protons, R_{CSDA}. By applying formula 1, $E_{incident}$ can be derived:

$$R_{CSDA}\left(E_{incident}\right) = R_{CSDA}\left(E_{DMG}\right) + x\rho, \qquad (1)$$

where x is the path length and ρ the density of the phantom, traversed by the incident proton beam before entering the DMG. If different materials are traversed before entering in DMG, formula 1 has to be applied to obtain the energy of the proton beam $E_{incident}$ at the entrance of each specific material.

4. Methodology

Geant4 based simulations [5] were developed to prove the principle of operation of DMG for QA in ocular and prostate proton therapy. The DMG was modelled accurately in terms of geometry and materials. A proton beam of 62 MeV and 129.46 MeV was adopted for ocular and prostate cancer treatment, respectively. In both cases the proton beam is incident along the major axis of the DMG. To describe particle interactions in the simulation toolkit, the Livermore Data Libraries were adopted to model the electromagnetic interactions, while the QGSP_BIC_HP Physics list was adopted to describe hadronic interactions. The threshold of production of secondary particles was fixed to 0.01 mm. The energy deposition produced by the incident proton beam and its secondary particles was recorded in the silicon strip detectors.

4.1. Simulation application for Ocular Cancer Therapy QA

In the case of pre-treatment QA in proton therapy for ocular melanoma, a 62 MeV proton beam was simulated with particles generated at -20 mm in air. A beam radius varying between 1 and 15 mm was adopted to study its effect on the response of the detector. The detector was positioned within a PMMA spherical phantom with radius equal to 18 mm, modelling a simplified anatomy of the eye while remaining large enough to house the detector as shown in figure 4.

4.2. Simulation Application for Prostate Cancer Therapy QA

A 129.46 MeV proton beam was selected to model the beam energy of the first experimental measurements with DMG, performed by the CMRP and collaborators at the Massachusetts General Hospital, Boston. In this simulation, a polystyrene target of varying thickness (70.9 mm, 81.1 mm and 91.3 mm) was positioned in front of the DMG with polystyrene scattering blocks placed above and below the DMG, as shown in figure 5.

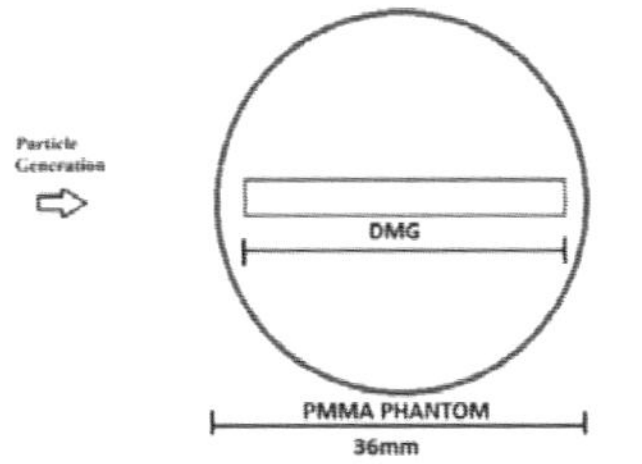

Figure 4. Simplified schematic of the simulation application setup for ocular cancer therapy QA.

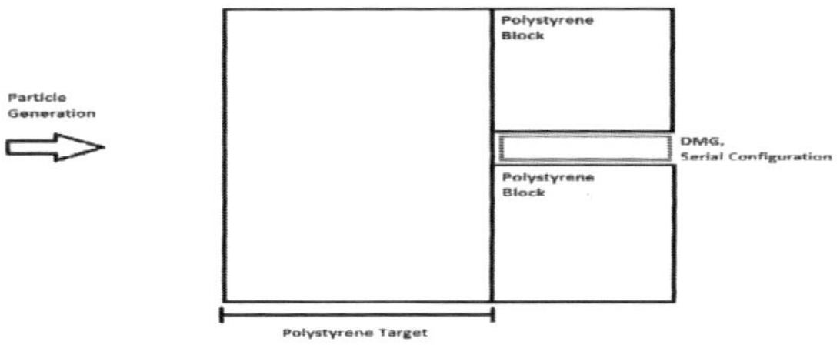

Figure 5. Simplified schematic of the simulation application setup for prostate cancer therapy QA.

5. Results

5.1. DMG for Ocular Cancer Therapy QA

Figure 6 shows the DMG response with an incident beam of 62 MeV protons, with different beam radius. The energy deposition is calculated in the silicon strip detectors and then converted to depth in the DMG.

Two distinct Bragg Peaks are observed in the detector response. As the beam radius increases, a greater number of protons scatter into the silicon sensitive strips, following the interactions in the surrounding PMMA phantom. This translates in a higher second Bragg Peak in DMG.

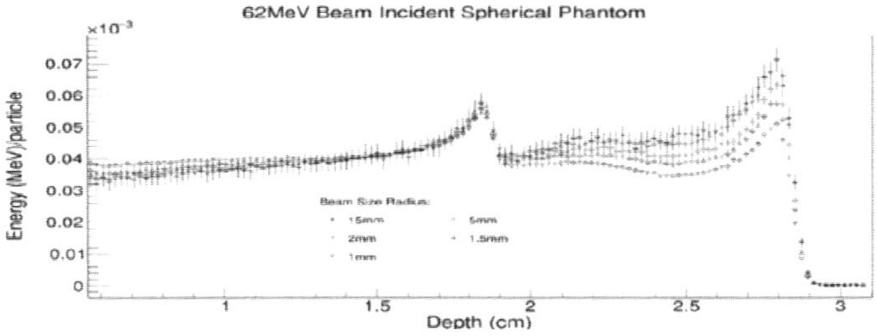

Figure 6. Response of the DMG to a mono-energetic 62MeV proton beam.

From the detector response, the distance between the two Bragg Peaks is determined to be 9.7 mm, corresponding to an energy of 60 MeV at the DMG entrance. Using the methodology outlined in section 3 the initial energy of the beam is determined to be approximately 62 MeV when entering in the phantom. This value agrees with the energy of the proton beam input to the simulation.

5.2. Application for Prostate Cancer Therapy QA

Response of the DMG in serial configuration to mono-energetic 129.46 MeV protons incident along the major axis of the detector is presented in figure 7 for the polystyrene thicknesses investigated. The range difference of Bragg Peaks is presented in Table 1 alongside with the calculated proton incident energy at the DMG entrance and the phantom entrance, derived using the method described in section 3. For polystyrene thickness of 70.9 mm, as the second peak occurs in the detector region of no response, analysis is unviable.

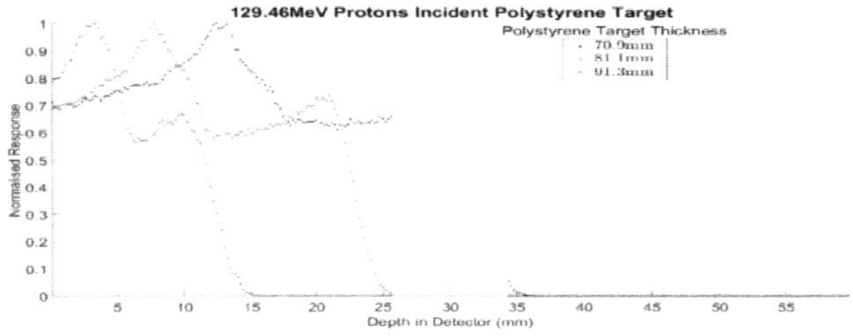

Figure 7. Response of the DMG in serial configuration to mono-energetic 129.46 MeV protons. In serial configuration a physical gap causes a region of no energy detection.

Table 1. Application for Prostate Cancer Therapy Results. E_{DMG} is the energy of the proton beam at the entrance of DMG. $E_{incident}$ is the energy of the proton beam at the entrance of the phantom.

Polystyrene Target Thickness (mm)	Peak Range Difference (mm)	E_{DMG} (MeV)	$E_{incident}$ (MeV)
81.1	13.2	61.25	131.6
91.3	6.4	41.25	128.9

6. Discussion and conclusions

The results of this work demonstrate the applicability of the DMG in pre-treatment QA to determine the energy of the incident proton beam.

The proof of principle has been demonstrated for ocular cancer therapy QA, with sharp, easily identifiable pristine Bragg Peaks observed in the detector response. This means that DMG in principle may be used to accurately resolve the initial beam energy, as demonstrated by the agreement in determined initial beam energy and injected particle energy.

When comparing the case of prostate cancer proton therapy against the treatment of ocular cancer, with a higher proton energy the position of the Bragg Peaks results to be less resolved as protons traverse a larger number of different materials surrounding the detector. This leads to a larger error in determining the distance between the Bragg Peaks measured by the DMG. Despite these effects, our study shows that the energy of the incident beam can be resolved in principle with an error of ~1 MeV for a mono-energetic proton beam.

In conclusion this work proved the principle of operation of DMG for pre-treatment QA in proton therapy for the determination of the proton beam energy. CMRP is currently undertaking experimental measurements with collaborators at the Massachusetts General Hospital to prove this principle of operation.

In the near future, the simulation results shown in this paper will be compared to the experimental measurements. The simulations will be repeated with a Spread Out Bragg Peak to verify the principle of operation in this case as well.

7. Acknowledgments

The author wishes to thank the financial support provided by Prince of Wales Hospital Radiation Oncology Trust Fund, without which research would not have been able to be undertaken.

8. References

[1] ICRU 2007 Prescribing, Recording and Reporting Proton-Beam Therapy (Report 78) *J. ICRU* **7**
[2] IAEA 2000 Absorbed Dose Determination in External Beam Therapy: An International Code of Practice for Dosimetry Based on Standards of Absorbed Dose to Water *Technical Report Series No. 398*
[3] Hill R F *et al* 2008 *Rad. Meas.* **43** 1258-64
[4] Wong J H D et al 2010 *Med. Phys.* **37** 427-39
[5] 2014 "GEANT4 User's Guide for Application Developers (Version: GEANT4 10.1)", GEANT4 Collaboration

Status of the ELIMED multidisciplinary and medical beam-line at ELI-Beamlines

F Romano[1], G A P Cirrone[1], G Cuttone[1], F Schillaci[1], V Scuderi[1,2], A Amico[1], G Candiano[1], S Giordanengo[2,3], L F Guarachi[2,3], G Korn[4], G Larosa[1], R Leanza[1,5], R Manna[1], V Marchese[1], F Marchetto[2], D Margarone[4], G Milluzzo[1,5], G Petringa[1,5], J Pipek[1], R Sacchi[2,3], A Vignati[2]
[1]Istituto Nazionale di Fisica Nucleare, Laboratori Nazionali del Sud, Via Santa Sofia 62, Catania, Italy
[2]Istituto Nazionale di Fisica Nucleare, Sezione di Torino, Via P. Giuria 1, Torino, Italy
[3]Universita' di Torino, Dipartimento di Fisica, Via P. Giuria 1, Torino, Italy
[4]Institute of Physics ASCR, v.v.i. (FZU), ELI-Beamlines Project, 182 21 Prague, Czech Republic
[5]Università di Catania,Dipartimento di Fisica e Astronomia,Via S. Sofia 64,Catania, Italy

E-mail: francesco.romano@lns.infn.it

Abstract. Nowadays, one of the biggest challenges consists in using high intensity laser-target interaction to generate high-energy ions for medical purposes, eventually replacing the old paradigm of acceleration characterized by huge and complex machines. In order to investigate the feasibility of using laser-driven ion beams for multidisciplinary application, a dedicated beam transport line will be installed at the ELI-Beamlines facility in Prague (CZ), as a part of the User-oriented ELIMAIA beam-line dedicated to ion acceleration and their potential applications. The beam-line section dedicated to transport and dosimetric endpoints is called ELIMED (ELI-Beamlines MEDical and multidisciplinary applications) and will be developed by the INFN-LNS.

1. Introduction

In the last few decades, charged particle acceleration using ultra-intense and ultra-short laser pulses has gathered a strong interest in the scientific community. Indeed, it could represent the future of particle acceleration and open new scenarios in multidisciplinary fields as, in particular, the medical one. Recently, high interest of the scientific community is driven from the fact that more compact laser-based therapy units could dramatically increase the availability of high-energy proton and carbon ion beams, and provide particle therapy to a broader range of patients [1, 2]. Furthermore, laser-based hadrontherapy would offer the advantage of having a unique integrated system, as for instance the possibility to perform treatments with different types of radiation sources: protons, heavier ions (He, Li, C), electrons, X/gamma rays and neutrons.

A variety of mechanisms underlying laser-driven ion acceleration exists dependant on the properties of the laser pulse and on the target geometry [3]. So far, most of the experimental results obtained refer to the TNSA regime [3], although features associated to several of the other mechanisms have recently emerged from experiments. Recently, it has been theoretically shown that PW-class lasers can accelerate protons to a few hundred MeV, thus fulfilling the medical requirements for particle energy. However, generation of an ion source that would simultaneously meet the maximum energy, energy spread, number of particles per bunch and repetition rate requirements for its application in hadrontherapy, is still a challenge. Laser-driven beams are, indeed, characterized by very intense (10^8-10^{12} particles per bunch) ultra-short (~ ps) particle pulses and ultra-high pulse dose rate (up to 10^{12} Gy/min) compared to conventional clinical proton beams (10^7-10^{10} particles/s and dose-rate up to 10-50 Gy/min) [3]. Besides, they have broad energy (up to 100% compared to 0.1-1%) and angular distributions and substantial intensity (charge) fluctuations from pulse to pulse. Thus, beyond standard requirements such as shot-to-shot operational stability, a reliable and precise dosimetric characterization of laser-accelerated particle beams is needed.

Several international collaborations and experiments have been launched in the last years and many research centers are currently involved in the investigation of laser driven therapy and applications and a more complete and extensive review on these research projects along with the specifications of the laser systems and technical approaches involved can be found in an ICFA Publication [4]. In this framework, a collaboration, named ELIMED, between the INFN-LNS (National Institute for Nuclear Physics - Laboratori Nazionali del Sud, Catania, Italy) and the ASCR-FZU (Institute of Physics of the Czech Academy of Science) has been established in 2011. The ELIMED project (ELI-Beamlines MEDical and multidisciplinary applications) aims to demonstrate the validity of new approaches based on laser-driven ion sources for potential future applications in medical and other multidisciplinary fields, including hadrontherapy.

2. The ELIMED beam-line

In 2018, a User-oriented beam-line, ELIMAIA (ELI Multidisciplinary Applications of laser-Ion Acceleration) equipped with diagnostics and dosimetry end-points will be commissioned at the ELI-Beamlines facility in the Czech Republic with the main goal to perform proof-of-principle experiments, dosimetry measurements and radiation biology investigations at high repetition rate. The INFN-LNS is in charge for the development, realization and installation of the beam-line section, called ELIMED from the collaboration, dedicated to the beam transport and energy selection, diagnostics and dosimetry of the laser-driven ion beams. The purpose is to provide to the scientific community interested in multidisciplinary and medical applications a stable controlled and reproducible beam. The ELIMED beam line consists of two main sections: the first one, in vacuum, dedicated to transport, collection and diagnostics of the accelerated beam and the second one, in air, addressed to the relative and absolute dose measurement [5]. Fig. 1 (left) shows a technical drawing of the whole ELIMED beamline. In particular, in the first section, a focusing quadrupole (PMQs) system [6] will be placed downstream the target allowing beam collection and focusing, followed by a resistive energy selector system (ESS) dedicated to the beam selection in terms of species and energy. The PMQs system consists of a 5 permanent magnet quadrupoles based on a hybrid Halbach scheme, with a 100 T/m gradient [6]. The radial field uniformity is better than 3%, while the longitudinal uniformity is better than 1%. The system will be placed downstream the target, it is modular and it can be adapted for the collection and optimization of reference beams with different energies (3-70 MeV protons) to be injected in the energy selection system. The energy selection system consists of 4 resistive C-shaped dipoles allowing the selection of a single reference trajectory in chicanes. Thanks to such layout the resolving power of the system depends only on the selection slit width (ranging between 1 and 20 mm to obtain energy spreads from 1 up to 20%). The resistive magnets will work with variable fields from 0.085 to 1.2 T ensuring the selection for C+6 up to 70 MeV/u and protons up to about 300 MeV.

In the in-air section of the ELIMED beam line, a dosimetric system composed by detectors for relative and absolute dosimetry will be installed (Fig. 1, right). Dose delivery with a 5% accuracy has to be ensured at the sample irradiation point; in order to fulfil this task, taking into account the extremely high dose rate per pulse, it is mandatory to design and realize dose-rate independent detectors. In particular, a secondary electron monitor (SEM) and a multi-gap ionization chamber (IC), dedicated to the on-line in-transmission relative, dosimetry will be realized. The multi-gap IC allows correcting for the ion recombination effects providing reliable dose measurements also for extremely variable intensities per pulse. These detectors will be calibrated using a Faraday cup (FC) for absolute dosimetry specifically designed to decrease uncertainties in the collected charge. After the calibration procedures, a sample irradiation system (SIS), placed at the end of the in-air section, will allow cell sample irradiations with a sub-millimetric precision.

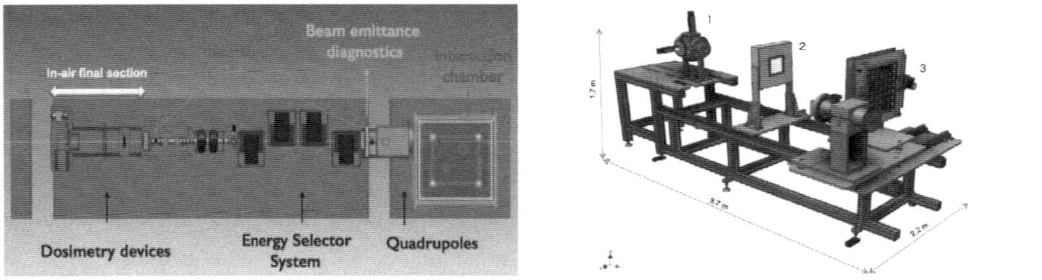

Figure 1. Left: schematic layout of the ELIMED beam line. Right: in-air section equipped with the SEM (1), the multi-gap IC (2) and, at the sample irradiation point, the FC and the SIS (3).

3. Geant4 Monte Carlo simulation of the ELIMED beamline and preliminary results

A complete Monte Carlo (MC) application of the ELIMED beamline has been developed with the Geant4 (GEometry ANd Tracking) toolkit [7]. Monte Carlo simulations have been widely used to support the design of some elements composing the beam line and to preliminary study the response of detectors. Moreover, once the final configuration of the beamline will be accurately reproduced, the Geant4 simulations will be used to predict the particle beam characteristics at specific positions along the beam line and to evaluate dose, fluence and particle distribution in the in-air section, where the irradiation will be performed.

Initially, the PMQs focusing system has been simulated using analytical transport codes, such as SIMION [8] and COMSOL [9], which allowed to design the geometry configuration and to fix the number of quadrupoles, their magnetic field features, their size and their optics dynamics. Then, the focusing system has been completely simulated also with Geant4, implementing the magnetic field maps created with the analytical transport software.

As discussed in the previous section, in order to deliver a controlled proton beam with energies up to 60 MeV/u and limit the energy spread, the focusing elements will be coupled with an innovative energy selection system. The energy selector system geometry has been fully reconstructed in the GEANT4 application and the magnetic field map grid of the single dipole, provided by COMSOL software, has been imported in the simulation.

For both the PMQ and the ESS the particle transport through the system has been simulated and compared with the results obtained with analytical transport softwares. The detailed results of such comparisons will be published elsewhere.

Once the final version of the ELIMED application has been validated and the whole beam line has been successfully implemented, in order to study the expected energy spectra, fluence and dose distributions, top-bottom simulations have been performed. These results aim to demonstrate the reliability of the simulated model, which now can be used to predict the necessary information at the different endpoints in the ELIMAIA beam line. The source input parameters have been extracted from PIC simulation corresponding to a typical TNSA spectrum with a maximum energy cut-off of about 100 MeV with an initial angular divergence ranging from about 40° half-angle for lowest energies and about 5° half-angle for the highest energy components. A preliminary simulation study to investigate the energy spreads and the transmission efficiencies achievable after the selection has been performed for different energies selected and for different selection slit widths. In particular, the energy spectra were registered after the ESS section, with all magnetic fields (i.e. quadrupoles, energy selector) configured to transport a given energy proton beam and the slit aperture set to 5 mm, 10 mm and 20 mm. Fig 2 (left) shows the energy spectra acquired just after the ESS varying the slit width using the beamline configuration properly tuned to transport 5 MeV protons.

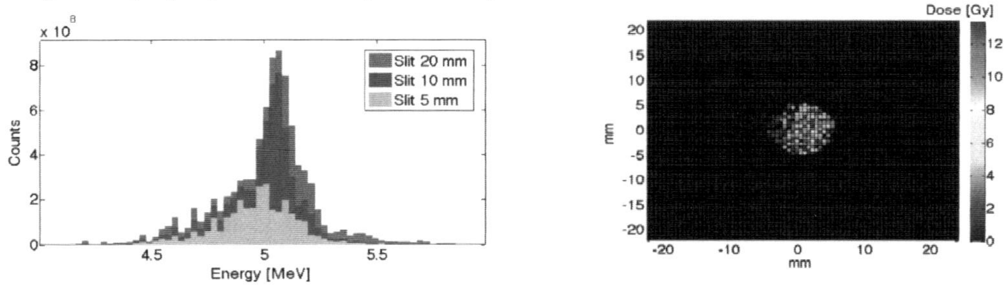

Figure 2. Left: energy spectrum for 5 MeV protons registered after the ESS with a 20 mm (blue), 10 mm (red) and 5 mm (yellow) slit aperture; the counts reported have been normalized on the real number of protons emitted obtained by the PIC. Right: spatial dose distribution in water for one shot.

The energy resolutions have been extracted from a gaussian fit of the energy spectra shown in figure 2. The energy resolution, obtained with the three slit configurations, ranges from 4 to 6 %. The transmission efficiencies, evaluated as the ratio of the transmitted (output) and the emitted (input) protons within the energy range 5 ± 0.25 MeV, are 2, 4.5 and 6% for 5, 10 and 20 mm slit aperture, respectively. The results indicate how the transmission efficiency strongly depends on the slit width and, as a consequence, a compromise for the slit width has to be found in order to assure a transmission of about 10% and at the same time a significant number of protons in the in air final section at the sample irradiation point. Moreover, a preliminary estimation of the 2D spatial dose distribution in water for one shot is shown in the same figure (right), demonstrating that sample irradiation with an acceptable beam homogenenity can be carried out.

A more extensive simulation study is currently on-going for the transport and selection of 30 MeV and 60 MeV proton beams; these energies are, indeed, more relevant to asses all the clinical aspects. In both cases, longitudinal transversal dose distributions at the irradiation point will be also retrieved from the simulations.

4. Acknowledgments

This work has been also supported by the ELI-Beamlines Con-305 tract no S 14-187, under Laser Gen (CZ.1.07/2.3.00/30.0057). This research was also supported by the Ministry of Education, Youth and Sports of the Czech Republic (ELI-Beamlines reg. no. CZ.1.05/1.1.00/02.0061).

5. References

[1] Bulanov S V *et al* 2014 *Physics-Uspekhi* **57** 1149-79
[2] Ledingham K W D *et al* 2014 *Appl. Sci.* **4** 402-43
[3] Macchi A *et al* 2012 *Reviews of Modern Physics* **85**
[4] Leemans W et al 2011 *ICFA Beam Dynamics Newsletters* **56** 51
[5] Cirrone G A P *et al* 2016 *Appl. Sci.* **5** (in-press. doi:10.3390/app50x000x)
[6] Schillaci F *et al* 2015 *J. Instrum.* **10** T12001
[7] Agostinelli S *et al* 2003 *Nucl. Instr. Meth. Phys. Res. A* **506** 250-303
[8] www.simion.com
[9] www.comsol.it

Patient specific quality control for Stereotactic Ablative Body Radiotherapy (SABR): it takes more than one phantom

T Kron[1-3], E Ungureanu[1], R Antony[1], N Hardcastle[3,4], N Clements[1], J Ukath[1], C Fox[1], P Lonski[1,2], D Wanigaratne[1], A Haworth[1,2]
[1]Peter MacCallum Cancer Centre, Department of Physical Sciences, Melbourne, VIC, Australia
[2]Peter MacCallum Institute, University of Melbourne, Parkville, VIC, Australia
[3]University of Wollongong, Centre for Medical Radiation Physics, NSW, Australia
[4]Royal North Shore Hospital, Sydney, NSW Australia

E-mail: tomas.kron@petermac.org

Abstract. Stereotactic Ablative Body Radiotherapy (SABR) is an extension of the concepts of Stereotactic Radiosurgery from intracranial procedures to extracranial targets. This brings with it new technological challenges for set-up of a SABR program and continuing quality assurance. Compared with intracranial procedures SABR requires consideration of motion and inhomogeneities and has to deal with a much larger variety of targets ranging from lung to liver, kidney and bone. To meet many of the challenges virtually all advances in modern radiotherapy, such as Intensity Modulated and Image Guided Radiation Therapy (IMRT and IGRT) are used. Considering the few fractions and high doses per fraction delivered to complex targets it is not surprising that patient specific quality control is considered essential for safe delivery. Given the variety of targets and clinical scenarios we employ different strategies for different patients to ensure that the most important aspects of the treatment are appropriately tested, be it steep dose gradients, inhomogeneities or the delivery of dose in the presence of motion. The current paper reviews the different approaches and phantoms utilised at Peter MacCallum Cancer Centre for SABR QA.

1. Introduction

Stereotactic Ablative Body Radiotherapy (SABR) is a relatively novel approach to cancer treatment where very high doses per fraction are given to small lesions using image guidance and motion management. SABR has become an accepted treatment modality for early stage lung cancer [1] and an increasing number of other curative and palliative indications [2]. Compared to conventional radiotherapy SABR features a number of challenges that require consideration for quality assurance:

- Doses exceeding 10Gy per fraction
- Small and possibly dynamic radiation fields
- Non-coplanar beam arrangements
- Inhomogeneity correction
- Assessment of motion during planning
- Motion management during treatment
- High quality and frequent image guidance

1.1 Typical SABR scenarios

Based on the intracranial experience, the 'classical' SABR scenario is a small lesion located in a large parallel-organised critical structure such as lung and liver. Compared to conventional radiotherapy SABR dose distributions are often not homogenous with high dose regions exceeding 125% of prescription dose being not uncommon. Tight margins around the target are normal as image guidance allows for target localisation directly prior to treatment [3-7]. An extension of this concept is intrafraction monitoring of tumour location, which can either be done with radiobeacons [8, 9] or intrafraction kilovoltage monitoring (KIM) [10].

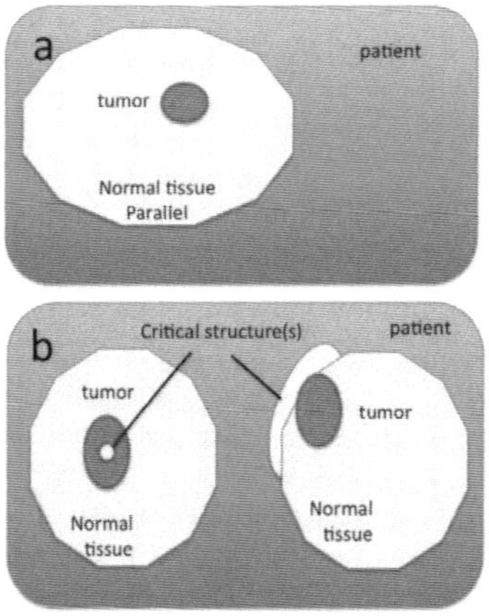

Figure 1. SABR scenarios a) 'classical' stereotactic problem, b) new SABR challenges

Figure 1 illustrates typical SABR scenarios. In the case of scenario a) the objective is to constrain the dose closely to the target and conformity indices become an important planning tool. The scenarios shown in figure 1b are emerging in clinical problems such as prostate SABR and the treatment of vertebral metastases on the left side of the figure and lesions close to intestines or other dose limiting structures on the right. Here steep dose gradients are important where for quality assurance spatial resolution becomes more important than absolute dose accuracy. Table 1 provides a summary of SABR scenarios and their specific challenges form a dosimetric point of view.

1.2. Patient specific quality assurance

Given these considerations SABR requires both machine and patient specific quality control (QC) activities to ensure SABR is delivered as planned. We report here on a suite of phantoms and QC approaches that were developed specifically for patient specific SABR QC at our institution.

2. Materials and Methods

A risk analysis was performed prior to commencing a SABR program for early stage lung cancer at Peter MacCallum Cancer Centre in 2009. As motion, small fields and inhomogeneity were considered key concerns, a Modus Quasar phantom which includes lung inhomogeneities was modified to allow for various motion patterns (figure 2) [11].

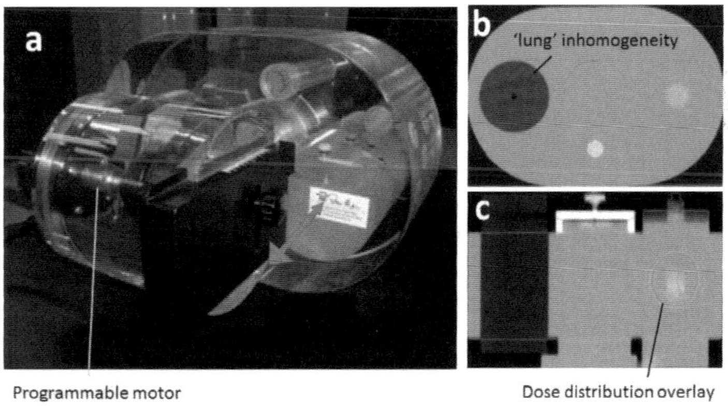

Figure 2. The modified QUASAR phantom (Modus Medical) featuring inhomogeneities and a programmable motor that allows mimicking customised motion patterns

The phantom was used for individual patient QC in our institution [12] as well as credentialing of a clinical trial of lung SABR (CHISEL, TROG 09.02). For the latter inhomogeneity correction was tested using small fields and the effect of motion was studied using radiochromic film [13]. It is well suited for assessment of dose distributions in coronal or sagittal plane as the cylinder holding dosimeters (right in figure 2b) can be rotated around a sup/inf axis. However, in some SABR applications the dose distribution in axial plane is required to demonstrate the steep dose gradients between target dose and spinal cord and oesophagus. For this purpose we designed a phantom to mimic the body of a patient with the particular aim to verify the steep dose gradients encountered in SABR of vertebral lesions. This phantom is shown in figure 3.

Table 1. SABR applications and patient specific quality control activities

Treatment scenario	Main challenges	Planning considerations	QC approach, Phantom/dosimetry
Lung: early stage curative intent	Inhomogeneity, motion, small fields	Energy <=10MV, Field size > 3 x 3cm^2, no VMAT FFF for single fraction	4DCT review, complex cases measure using QUASAR (Figure 2)
Lung: oligometastases	Inhomogeneity, motion, multiple lesions	As above, limitations on non-coplanar to limit overlap	As above
Liver	Motion, contrast CT, IGRT	Gating or breath hold considered	4DCT review, IGRT strategy review
Kidney	Motion, skin dose, IGRT	Consider 18MV, non-coplanar approach	4DCT review, in vivo dosimetry for skin
Bone metastases	Irregular size, close to skin (eg sternum)	Consider electron contribution	Consider in vivo dosimetry for skin
Vertebral body	Steep dose gradients required	IMRT > 9 fields VMAT	Phantom measurement using Rod (Figure 3)
Prostate	Irregular motion, urethra in centre	Fiducials common, spacers, rectal balloons	On-line imaging

Additional SABR Quality Control activities for individual patients depend on the scenario. Some important ones are listed in table 1. In addition to this we can utilise an independent dose calculation tool, Mobius 3D and FX. The latter relies on MLC dynalog files to determine dose distribution in the planning scan using a superposition convolution algorithm [14].

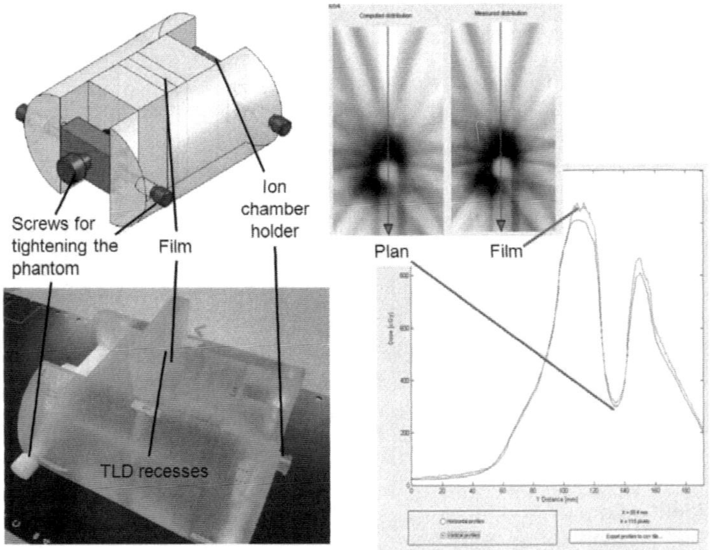

Figure 3. The 'Rod' phantom for SABR spine QA measurements. The evaluation relies largely on radiochromic film but can also include ion chambers and TLD measurements.

3. Results

While the introduction of motion did not alter the QC results in the phantom, the introduction of inhomogeneities as shown in figure 2, did. The initial measurements for individual patients confirmed that the Analytical Anisotropic Algorithm (AAA) used in the Varian Eclipse planning system has difficulties to predict the dose behind inhomogeneities accurately [15, 16]. After the initial phase of measurements this is now only verified using an independent monitor unit calculation and general checks of the treatment planning system.

Given the fact that 3D conformal treatments could in general be verified without any significant problems once the treatment couch had been taken into consideration, patient specific QC activities have been reduced [12]. However, in the context of SABR for vertebral lesions, where Intensity Modulated Radiation Therapy is essential, every patient treatment is still verified using physical measurements prior to treatment. For this an ionisation chamber measurement and a radiochromic film assessment in the "Rod" phantom shown in figure 3 is routinely performed. It is particularly the dose distribution recorded on the film as shown in figure 3 on the right which informs the acceptability of the plan. The steep dose gradient in both plan and treatment verification film can be clearly seen and a 1mm distance to agreement criterion is typically used for evaluation.

4. Discussion and Conclusion

For the wide variety of SABR applications no single phantom appears to be suitable for individual patient QC as different treatment approaches require visualisation of dose distributions in different planes with high spatial resolution. In addition to ionisation chambers measurements radiochromic film was found to be essential for most of the measurements. Future work will be directed to studying real time dosimeters with high spatial resolution such as the dose magnifying glass [17] as replacement to shorten the turn around time for QC measurements.

5. References

[1] Timmerman R D *et al* 2010 *J. Clin. Oncol.* **28** 907-9
[2] Timmerman R D *et al* 2007 *J. Clin. Oncol.* **25** 947-52
[3] Dawson L A and Jaffray D A 2007 *J. Clin. Oncol.* **25** 938-46
[4] Korreman S *et al* 2010 *Radiother. Oncol.* **94** 129-44
[5] Vial P *et al* 2008 *Med. Phys.* **35** 1267-77
[6] Hill R *et al* 2010 *Med. Phys.* **37** 4355-63
[7] Greer P B *et al* 2007 *Med. Phys.* **34** 4389-98
[8] Kupelian P *et al* 2007 *Int. J. Radiat. Oncol. Biol. Phys.* **67** 1088-98
[9] Poulsen P R *et al* 2015 *Acta. Oncol.* **54** 1445-52
[10] Keall P J *et al* 2015 *Med. Phys.* **42** 354-8
[11] Dunn L *et al* 2012 *Australas. Phys. Eng. Sci. Med.* **35** 93-100
[12] Hardcastle N *et al* 2014 *Australas. Phys. Eng. Sci. Med.* **37** 45-52
[13] Kron T *et al* 2011 *Radiat. Meas.* **46** 1920-23
[14] Kerns J R *et al* 2014 *Radiat. Oncol.* **9** 176
[15] Dunn L *et al* 2015 *Phys. Medica.* **31** 435-41
[16] Van Esch *et al* 2006 *Med. Phys.* **33** 4130-48
[17] Wong J H *et al* 2010 *Med. Phys.* **37** 427-39

Experimental studies with two novel silicon detectors for the development of time-of-flight spectrometry of laser-accelerated proton beams

M Würl[1,4], S Reinhardt[1,5], A Rosenfeld[2], M Petasecca[2], M Lerch[2], LTran[2], S Karsch[3], W Assmann[1], J Schreiber[1], K Parodi[1]
[1]Department of Medical Physics, Faculty of Physics, Ludwig-Maximilians-Universität München, Am Coulombwall 1, 85748 Garching b. München, Germany
[2]Centre for Medical Radiation Physics, University of Wollongong, Wollongong, NSW 2522, Australia
[3]Department for Laser Physics, Faculty of Physics, Ludwig-Maximilians-Universität München, Am Coulombwall 1, 85748 Garching b. München, Germany

E-mail: matthias.wuerl@physik.uni-muenchen.de

Abstract. Laser-accelerated proton beams exhibit remarkably different beam characteristics as compared to conventionally accelerated ion beams. About 10^5 to 10^7 particles per MeV and msr are accelerated quasi-instantaneously within about 1 ps. The resulting energy spectrum typically shows an exponentially decaying distribution. Our planned approach to determine the energy spectrum of the particles generated in each pulse is to exploit the time-of-flight (TOF) difference of protons with different kinetic energies at 1 m distance from the laser-target interaction. This requires fast and sensitive detectors. We therefore tested two prototype silicon detectors, developed at the Centre for Medical Radiation Physics at the University of Wollongong with a current amplifier, regarding their suitability for TOF-spectrometry in terms of sensitivity and timing properties. For the latter, we illuminated the detectors with short laser pulses, measured the signal current and compared it to the signal of a fast photodiode. The comparison revealed that the timing properties of both prototypes are not yet sufficient for our purpose. In contrast, our results regarding the detectors' sensitivity are promising. The lowest detectable proton flux at 10 MeV was found to be 25 protons per ns on the detector. With this sensitivity and with a smaller pixelation of the detectors, the timing properties can be improved for new prototypes, making them potential candidates for TOF-spectrometry of laser-accelerated particle beams.

1. Introduction
With the advent of petawatt laser systems, laser-driven ion acceleration may become an alternative way of producing ions with energies relevant for radiation therapy or medical imaging [1]. However, characteristics of laser-accelerated particle beams are quite unique and remarkably different compared to conventionally accelerated particle beams. They typically exhibit high divergence, a small source size and a wide spectral bandwidth [2]. About 10^5 to 10^7 particles per MeV and msr, depending on the resulting kinetic energy are accelerated quasi-instantaneously within about 1 ps [3]. Due to relatively

[4] Corresponding author
[5] Present address: Deutsches Patent- und Markenamt, Zweibrückenstr. 12, 80331 Munich, Germany

large shot-to-shot fluctuations in the energy spectrum and other physical properties of the ion bunch, a precise characterization of each particle pulse is required for practical applications. This information is often obtained using spectrometers based on electric or magnetic deflection. A convenient alternative would be to exploit the time-of-flight (TOF) difference of ions with different kinetic energies in order to measure the energy spectrum of the beam and its spatial distribution. A crucial requirement for this approach is fast and sensitive detectors, as well as broadband amplifiers and readout. The rise time of the detector determines the accuracy of the energy spectrum measurement, while the signal fall time of the system needs to be short ensuring that the low-energetic part of the proton spectrum can still be resolved and is not hidden in a saturated signal.

We tested two prototype silicon detectors, developed at the Centre for Medical Radiation Physics at the University of Wollongong regarding their suitability for TOF-spectrometry of laser-accelerated proton pulses. Our interest was to examine the timing properties and to estimate the lowest detectable proton number, since this puts a lower boundary for the size of pixelation.

2. Experimental setup

2.1. Detector prototypes

Two prototype silicon detectors were tested regarding their suitability for TOF measurements of laser-accelerated proton spectra. The first detector ("bridge detector") is a pseudo-pixelated detector with a sensitive thickness of 10 μm. The total chip size is 3.6 × 4.1 mm and it consists of three segments, each with two arrays of small pixels connected in parallel in each array. The pixel size is 30 x 30 μm^2.

The sensitive thickness of the second detector ("thin detector") is also 10 μm, but it was manufactured without supporting wafer. This makes it a potential candidate to be used in transmission mode to obtain the proton spectrum prior to an interaction with the final target for a specific application. Although the prototype we tested is not pixelated, segmentation of this detector in order to gain also spatial information about the proton spectrum, is foreseen.

2.2. Setup for the determination of the rise and fall times of the detectors

The timing properties of the prototype detectors were evaluated using short light pulses from the ATLAS Ti:Sapphire laser system at the Laboratory for Extreme Photonics (LEX Photonics) in Garching near Munich, Germany. The pulse duration at the detector position was well below 100 fs FWHM and it was centered at 795 nm wavelength. Several absorptive neutral density filters were placed prior to the detector, thus the pulse energy at the detector was in the order of tens of μJ.

Both detectors were reverse biased, assuring full depletion, and were coupled to a high-speed broadband current amplifier *(HSA-X-2-20, FEMTO, Germany)*. The amplified signal was visualized in a fast oscilloscope with a bandwidth of 4 GHz *(WaveRunner 640Zi, Teledyne LeCroy, USA)*, using the 50 Ω DC input. A commercial photodetector *(DET10A/M, Thorlabs, USA)* with a rise time of about 1 ns was used as a reference and for triggering the oscilloscope. It was placed adjacent to the detectors in order to measure the scattered laser pulses.

2.3. Setup for estimating the lowest detectable proton flux

The aim of this estimation is to judge whether the expected proton fluence from laser-acceleration at our facility will be sufficient to generate a signal in the detector which is well above noise level. We mimicked the energy deposition of a proton beam in the thin detector by illuminating it with short light pulses from a frequency-doubled Nd:YAG laser (λ = 532 nm). The pulse width was about 15 ns FWHM and the repetition rate was 10 Hz. We used the same setup as described in section 2.2. The laser intensity was then gradually reduced by adding absorptive neutral density filters until the measured photocurrent only scarcely exceeded the noise level. The current amplifier was then replaced by a charge sensitive pre-amplifier *(model PCP 5, FAST ComTec, Germany)*, followed by a spectroscopy amplifier *(model 2021, Canberra Industries, USA)* and a multichannel analyzer *(MCA8000A, Amptek, USA)* which was connected to a PC. To avoid saturation when measuring the produced integral signal charge, we used an additional neutral density filter which reduces the laser intensity by a factor of 100, as compared to the experiments using the current amplifier. Prior to the experiments at the Nd:YAG laser, we performed an energy calibration of the detector using a mixed alpha source, consisting of Am-241, Cm-244 and Pu-239.

2.4. Experimental setup at the Munich Tandem van-de-Graaf accelerator

First experiments with the thin detector have been performed at the Tandem van-de-Graaff accelerator of the Maier-Leibnitz-Laboratorium (MLL) in Garching. Protons were accelerated to an energy of 20 MeV. A chopper was used at the low energy side of the accelerator to obtain short pulses of about 120 ns. From the beam current which was measured using a Faraday Cup, we could estimate the proton flux on the detector. Within each pulse, the proton flux at the detector was about 6×10^3 protons/cm^2/ns and could be reduced by a factor of 30 by using an additional attenuator. The proton energy could be lowered to around 14 MeV by placing a 1 mm thin aluminum plate between vacuum window and detector. We used the same amplifier and read-out as described in section 2.2.

3. Results and Discussion

3.1. Timing properties of the detector prototypes

The photocurrent measured with the photodiode is compared to the current measured by the bridge detector and the thin detector (figure 1). The high-frequency noise, which was induced by the electronics of the laser system, could only partially be suppressed. For the bridge detector, the signal rise time from 10% to 90% was found to be 0.42 ± 0.02 ns, which is comparable to the measured rise time of the fast photodiode. The time it took for the signal to recover from the 90% to the 10%-level was 2.88 ± 0.36 ns, which is slightly longer than for the photodiode. The relatively long tail of the signal decay can be attributed to electron-hole pairs created in the non-depleted silicon under the bridge connecting two adjacent pixels. Timing properties of the thin detector were found to be poorer. We measured a rise time of 2.76 ± 0.34 ns and a fall time around 20 ns.

For a reasonable accuracy of the TOF spectrometry of laser-accelerated proton pulses, sub-ns time resolution is required. This was achieved only by the bridge detector. However, its fall time is rather long which will result in a loss of sensitivity due to saturation, when it comes to the TOF spectrometry of the low-energetic part of proton energy spectra. Since we assume that the insufficient timing properties of the thin detector is mostly attributed to its capacitance, we are aiming to reduce it for the next prototype of the thin detector by pixelating it.

3.2. Estimation of the lowest detectable proton flux

Protons with a kinetic energy of 10 MeV will deposit on average 81 keV in the 10 μm sensitive thickness of the detector [4]. According to the energy calibration using the mixed alpha source, the deposited energy in the thin detector was 0.3 MeV per laser pulse, i.e., the total deposited energy of a laser pulse which scarcely exceeds the noise level in current mode is 30 MeV. Since the laser pulse duration was 15 ns FWHM, the lowest detectable flux for 10 MeV protons is 25 protons per ns and sensitive detector area. For higher proton energies, the required proton flux slightly increases due to their lower energy deposition in the detector.

Realistic proton yields from laser acceleration experiments at 10 MeV are around 10^5 to 10^6 per MeV, msr and pulse [3]. In a distance of 1 m, which is a convenient drift space for TOF spectrometry at our facility, the resulting proton flux would thus range between 10^2 and 10^3 mm^{-2} ns^{-1}. The expected proton flux is therefore sufficient to be detected when operating the thin detector in current mode. For the next prototype of this detector, segmentation is foreseen. Based on our investigations, a sensitive area of each segment of around 1 mm^2 is planned.

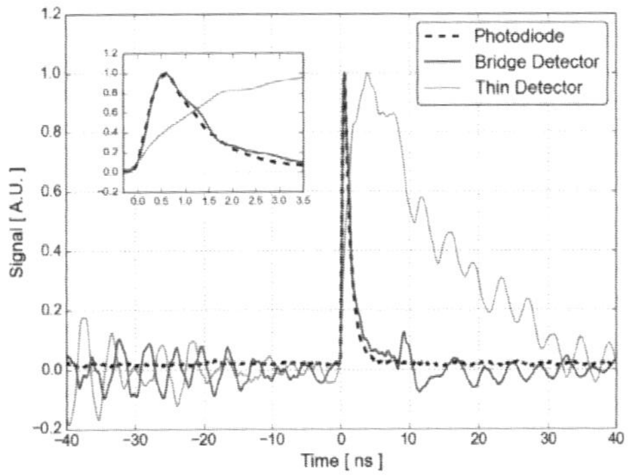

Figure 1. Photodiode signal compared to signal from bridge detector and thin detector. The inset shows a zoom in the time axis.

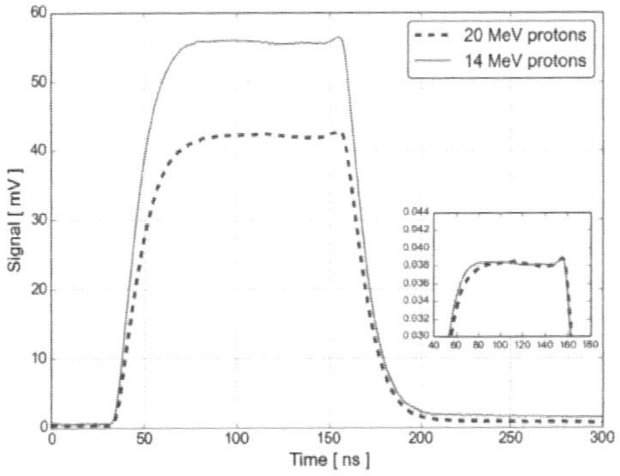

Figure 2. Current signal (converted to voltage) of mono-energetic proton pulses. In the inset, the signal is normalized by the proton number and the average energy deposition in 10 µm silicon for the respective proton energy.

3.3. Experiments at the Munich Tandem van-de-Graaf accelerator

In the experiments at the local Tandem accelerator we obtained a clear current signal for the attenuated proton pulses (figure 2), where according to beam current measurements, the fluence on the detector was around 200 protons per ns. The ratio of the signal height of proton pulses with 20 MeV and 14 MeV perfectly fits the ratio of the corresponding energy deposition in 10 µm silicon (inset in figure 2). In case of the non-attenuated proton beam with a fluence of around 6000 proton per ns, saturation of the detector could already be observed. However, no detailed study on the saturation limit has been performed so far. Yet, this problem can be solved by pixelation of the detector in the next prototypes.

4. Conclusion

We tested two prototype silicon detectors regarding their suitability for TOF-spectrometry of laser-accelerated proton beams in terms of timing properties and sensitivity. By using a broadband current amplifier and a fast oscilloscope the generated signal current in each detector was visualized.

According to our experiments, the timing properties of both detector prototypes are not yet sufficient for quantitative TOF-spectrometry. While the relatively long fall time of the bridge detector is attributed to its specific design, the long rise and fall times of the thin detector can most likely be improved by a lower capacitance. This can be achieved by pixelation, which will also increase the detector's saturation limit. On the other hand, from the signal height we could estimate that at 10 MeV at least 25 protons per ns on the detector are required to generate a signal that clearly exceeds the noise level. This hence gives a lower boundary for the pixel size.

5. Acknowledgements

We would like to thank J. Wenz, K. Khrennikov, H. Ding and C. Kreuzer for their help setting up the lasers. This work was supported by the DFG Cluster of Excellence Munich Centre for Advanced Photonics (MAP). M. Würl also acknowledges financial support from IMPRS-APS.

6. References

[1] Malka V *et al* 2008 *Nat. Phys.* **4** 447
[2] Daido H *et al* 2012 *Rep. Prog. Phys.* **75** 056401
[3] Henig A *et al* 2009 *Phys. Rev. Lett.* **103** 245003
[4] Berger M, Hubbel J, Seltzer S, Chang J, Coursey J, Sukumar R, Zucker D and Olsen K 2005 [Online] Available: http://physics.nist.gov/Star [2016, Jan 12]. NIST, Gaithersburg, MD.

IOP Conf. Series: Journal of Physics: Conf. Series **777** (2017) 012019 doi:10.1088/1742-6596/777/1/012019

New silicon microdosimetry probes for RBE and biological dose studies using stationary and movable targets in ^{12}C ion therapy

L Chartier[1], L T Tran[1], D Bolst[1], A Pogossov[1], S Guatelli[1], M Petasecca[1], M Lerch[1], D Prokopovich[1,2], M Reinhard[2], V Perevertaylo[3], M Jackson[4], N Matsufuji[5], A B Rosenfeld[1]

[1]Centre for Medical Radiation Physics, University of Wollongong, Wollongong, Australia

[2]Institute of Materials Engineering, Australian Nuclear Science and Technology Organization, Lucas Heights, Australia

[3]SBA-BIT, Kiev, Ukraine

[4]University of New South Wales, Sydney, Australia

[5]Research Centre for Charged Particle Therapy, National Institute of Radiological Sciences, Japan

E-mail: lc752@uowmail.edu.au

Abstract. Due to the high LET and dense ionisation tracks associated with ions, microdosimetric approaches have been used in carbon ion therapy to assess field quality and calculate radiobiological quantities for a variety of cell lines. There is however a lack of instrumentation for simple and routine use in a clinical environment, important for determination of RBE which provides accurate treatment planning and delivery in hadron therapy. In this study, a 10 μm thick silicon microdosimeter with 3D sensitive volumes has been used to investigate the effect of motion on the RBE and field quality of a typical ^{12}C ion therapy beam. For a passively scattered 290 MeV/u ^{12}C beam with 6 cm spread-out Bragg peak (SOBP), variations in biological dose along the SOBP were observed, as well as a significant changes to particle LET when incident on a moving target.

1. Introduction

In heavy ion therapy it is of paramount importance to ensure that the biological dose calculated in the treatment plan is identical to that given to the patient. There is however a lack of instrumentation for simple and routine use in a clinical environment, important for determination of RBE which provides accurate treatment planning and delivery in hadron therapy. Whilst preliminary theoretical calculations and cell studies have been undertaken [1], the effect that patient and organ motion has on these quantities is an issue that needs further investigation, especially with advancing technology such as proton and ^{12}C pencil beam scanning systems. Microdosimetry is a useful technique for RBE determination in unknown and mixed radiation fields typical of hadron therapy, currently implemented in Japan [2, 3]. However, the gold standard microdosimeter, the tissue equivalent proportional counter (TEPC), has a sensitive volume diameter that is larger than the typical motion amplitude, rendering it ineffective at detecting motion and misalignment anomalies. Microdosimeters with high spatial resolution are required to fully understand these effects. In this study, a 10 μm thick silicon microdosimeter with 3D sensitive volumes has been used to investigate a typical ^{12}C ion therapy beam, as well as the effect of motion on RBE and field quality.

2. Method

The detector used in this study is the Bridge Microdosimeter, designed by the Centre for Medical Radiation Physics (CMRP), University of Wollongong, Australia. The microdosimeter is an n-type silicon-on-insulator (SOI) detector based on an array of 30 x 30 µm sensitive volumes (SVs) with a thickness of just 10 µm, achieved through the use of masked ion plasma etching. Silicon "bridges" are left unetched to support the aluminium tracks between each SV. The microdosimeter is connected to an in-house designed low noise optimised preamplifier and shaping amplifier, enabling detection of lineal energies as low as ~0.2 keV/µm. The charge collection in this device has been characterised previously using the ion beam induced charge collection technique at ANSTO [4]. A high resolution scanning electron microscope (SEM) image of the Bridge microdosimeter showing the parallelepiped SVs and an image of the "MicroPlus" probe are pictured in figure 1.

Figure 1. The MicroPlus probe with an inset SEM image of the Bridge Microdosimeter sensitive volumes.

Previous studies with the Bridge microdosimeter have used a modular PMMA phantom to adjust the detector depth [4, 5], however this setup is relatively inflexible due to the set thickness of PMMA slabs. To overcome this limitation, a relatively low-cost XY-stage was built using a MakerBot Replicator 3D-printer and PMMA water tank, controlled remotely using Arduino-driven stepper motors. Not only does such a system allow flexibility in controlling the depth and lateral position of the microdosimeter, it also facilitates custom movements, used in this study to simulate simple linear motion as well as lung motion. This is achieved by moving the depth and lateral position of the detector using predefined profiles, shown in figure 2.

Experiments were performed at the Heavy Ion Medical Accelerator in Chiba (HIMAC), NIRS, Japan, utilising a passively scattered ^{12}C ion beam to irradiate the microdosimeter. Using a ridge filter designed to modulate the beam, multiple pristine Bragg peaks of discrete energies were created with a maximum energy of 290 MeV/u. The resultant field had a 6 cm spread-out Bragg peak (SOBP) with the assumption of constant biological dose along its plateau. The biological dose along the SOBP was calculated using measurements taken with the microdosimeter (RBE_{10} derived from microdosimetric spectra) and a PTW pinpoint ionization chamber (physical dose). The effect of lung motion was also studied close to the end of the SOBP along the beam axis, important for cases when a critical organ may be irradiated by high LET particles.

A spherical polyethylene bolus used in passive beam treatments to produce distal conformity was then placed upstream along the central axis of the beam. The microdosimeter was placed at various off-axis positions in the tank, labelled A and B in figure 3. Microdosimetric spectra were obtained while the microdosimeter was stationary at each position, in addition to constant motion from 20 mm to 40 mm off-axis, towards the edge of the bolus.

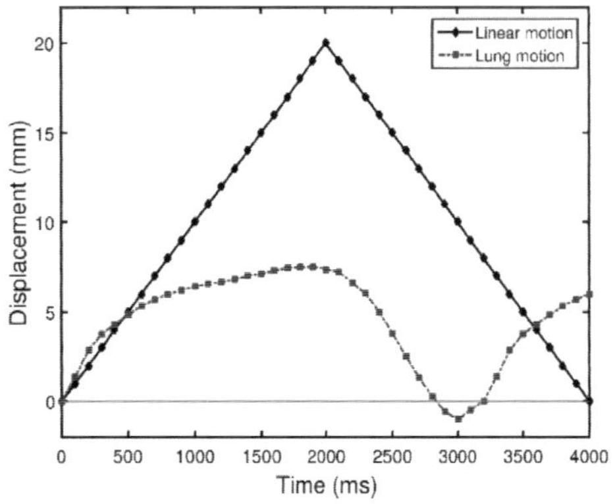

Figure 2. Linear and lung motion profiles

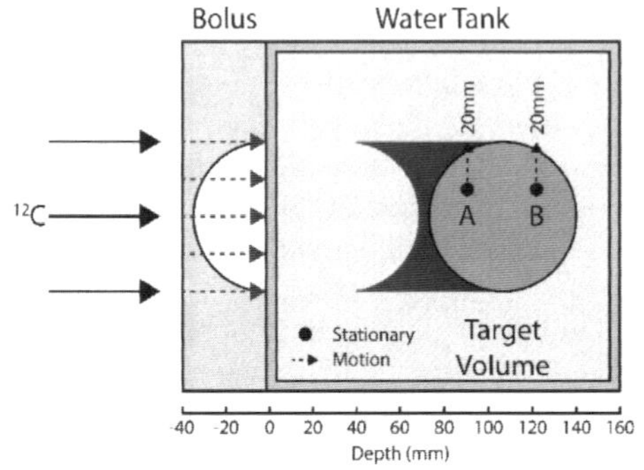

Figure 3. Schematic showing the phantom-bolus setup with acquisition points (A, B), and motion vectors.

3. Results

RBE_{10} was derived from the microdosimetric spectra and physical dose measurements were carried out with the ionisation chamber over the depth of the phantom, shown in figure 4. These distributions were found to be typical of heavy ion SOBP fields where, in order to produce constant biological dose in the target volume, the physical dose is made to decrease with depth due to increasing RBE_{10}. However, from figure 5, convolving these quantities showed that the biological dose along the SOBP is only relatively constant up until 120 mm, after which it decreases slightly then sharply increases at the distal part of the SOBP. Due to the high spatial resolution of the microdosimeter compared to TEPCs in previous studies [3], a more detailed biological dose distribution was measured.

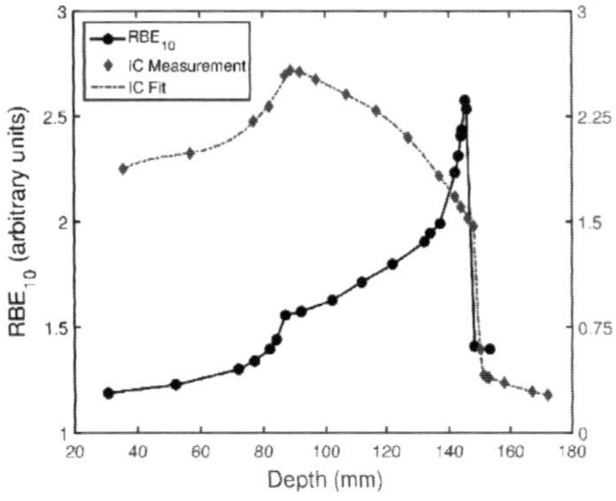

Figure 4. Charge collected from the ionisation chamber and RBE_{10} derived from microdosimetric spectra as functions of depth in water.

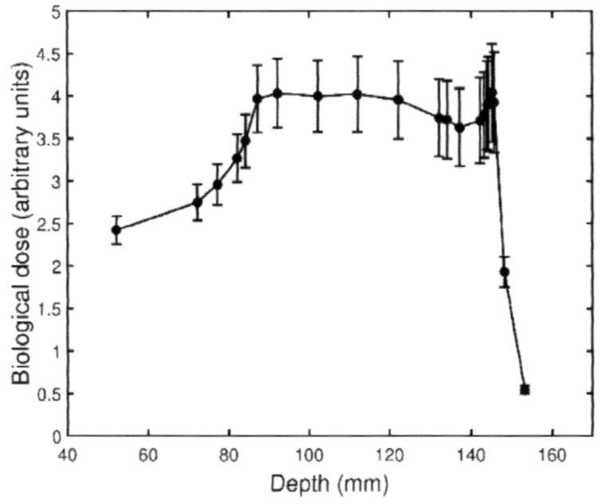

Figure 5. Biological dose distribution along the depth of the water phantom.

A significant difference was observed between the stationary spectra at a depth of 148mm and when undergoing lung motion from 141-149mm, seen in figure 6. When in motion, the microdosimeter enters the highest LET region of the field, resulting in lineal energy events from carbon ions dominating the spectra. In the stationary case, lower lineal energy events from neutron interactions dominate. From

IOP Conf. Series: Journal of Physics: Conf. Series **777** (2017) 012019 doi:10.1088/1742-6596/777/1/012019

this observable change in field quality, significantly more damage would be delivered to the non-target volume in a clinical scenario, as expected.

When undergoing motion from stationary points A and B (see figure 3), it can be seen from Fig. 7 that the spectra in both cases shifts towards higher lineal energies due to the contribution of high LET carbon ions stopping at the end of their range, towards the edge of the bolus. An increase in low lineal energy events was also observed when in motion from point B due to a higher contribution of fragments and neutrons produced from particle interactions further upstream.

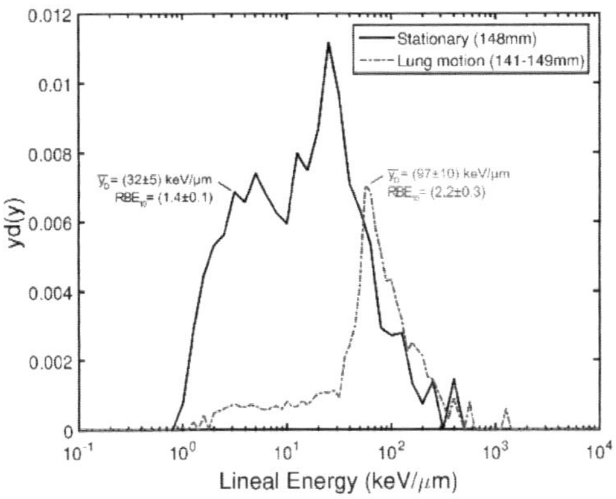

Figure 6. Microdosimetric spectra from stationary and moving acquisitions at points A and B with the bolus.

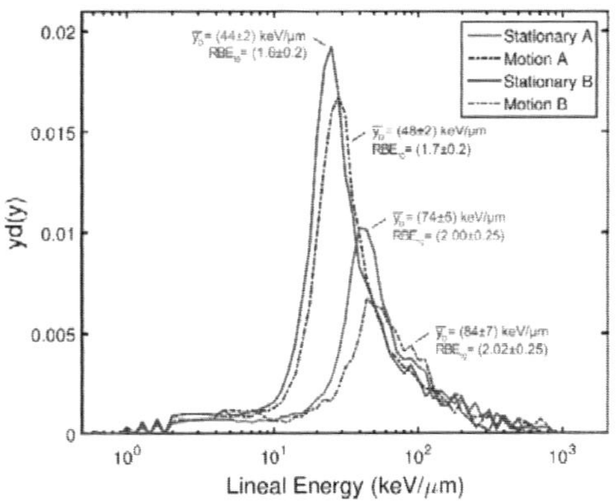

Figure 7. Microdosimetric spectra from stationary and moving acquisitions from depths 148mm, and undergoing lung motion from 141-149mm.

4. Conclusion

The CMRP has developed and implemented a new microdosimeter probe, enabling measurement of microdosimetric spectra with lineal energies as low as ~0.2 keV/µm. We have demonstrated that motion can lead to significant changes in the microdosimetric spectrum, and consequently the RBE, resulting in undesirable treatment outcomes. Due to the high spatial resolution of the microdosimeter compared to gold standard tissue equivalent proportional counters, a more detailed distribution of biological dose and other quantities can be measured, paving the way for routine QA procedures using solid-state microdosimetry.

5. References

[1] Gemmel A *et al* 2011 *Phys. Med. Biol.* **56** 7337
[2] Kase Y *et al* 2006 *J. Radiat. Res.* **166** 629
[3] Kase Y *et al* 2013 *J. Radiat Res.* **52** 59
[4] Tran LT et al 2015 *IEEE. Trans. Nucl. Sci.* **62** 504
[5] Tran LT et al 2015 *IEEE. Trans. Nucl. Sci.* **62** 3027

The angular dependence of a two dimensional monolithic detector array for dosimetry in small radiation fields

N Stansook[1], M Petasecca[1,2], K Utitsarn[1], M Newall[1], P Metcalfe[1,2], M Carolan[3,2], M Lerch[1,2], A B Rosenfeld[1,2]
[1]Centre for Medical Radiation Physics, University of Wollongong, NSW 2500, Australia
[2]Illawara Heath Medical Research Institute, Wollongong, NSW 2522, Australia
[3]Illawarra Cancer Care Center, Wollongong NSW 2500, Australia

E-mail: ns938@uowmail.edu.au

Abstract. The purpose of this study is to investigate the directional dependence of a two dimensional monolithic detector array (M512) under 6 MV photon irradiation and to evaluate the effect of field size on angular dependence. Square fields of sizes: 3x3 cm^2 and 10x10 cm^2 were measured at the iso-centre of a cylindrical phantom. Beam angles with incidences from 0^0-180^0 in increments of 15^0 were used to investigate the central pixel angular response of M512, normalized to the pixel response for normal (0°) beam incidence. The angular response of the detector was compared to the response of EBT3 radiochromic film in the identical geometric orientation. The maximum angular dependence was observed at the angle 90°±15° to be -18.62% and -17.70% for the field sizes 3x3 cm^2 and 10x10 cm^2, respectively. The angular dependence of M512 showed no significant difference between field sizes of 3x3 cm^2 and 10x10 cm^2 ($p>0.05$). The maximum dose difference measured by the central pixel of M512 and EBT3 for all angles are -20% for 3x3 cm^2 field size and -18.58% for the 10x10 cm^2 field. The diode array's size and packaging effects the angular response of the detector. The angular correction factor is necessary to apply to increase accuracy in dosimetry for arc treatment delivery.

1. Introduction

Advanced external beam radiotherapy (EBRT) techniques such as stereotactic body radiotherapy (SBRT) and volumetric modulated arc therapy (VMAT) have been increasingly used in the treatment of cancer patients. Due to their delivery complexity, with small beamlets and high dose gradients modulated throughout the rotation of the gantry around the patient, the accuracy of patient verification before treatment becomes very important. Small field dosimetry introduces significant complications for simulations and measurement because of electronic disequilibrium and other factors [1]. The ideal dosimeter for the verification of a dose distribution requires a small sensitive volume, high spatial (submillimetre) resolution, energy independence, dose rate independence and tissue equivalence [2-5].

The diode is a widely used radiation detector for small field dosimetry due to their suitable characteristics such as high sensitivity compared with the ionization chamber, small active volume size, high spatial resolution and real time measurement. However, the limitations of the diode include: energy dependence, dose rate dependence, and angular dependence [6]. Paul et al [7] studied the directional response of the surface p-Si type diode, they found the sensitivity varies within ±12% for incident beam angles from 0°-180°. Due to the effects of detector configuration and materials, similar trends in the

IOP Conf. Series: Journal of Physics:Conf. Series **777** (2017) 012020 doi:10.1088/1742-6596/777/1/01202(

angular response of n-Si type diode arrays has been reported. The response of MapCHECK (SunNuclear, Melbourne, Florida-USA) showed significant difference of up to 20-30% at the incident beam angle of 90° for a 6 MV photon field when compared with a cylindrical ionization chamber [7, 8]. The Centre for Medical Radiation Physics (CMRP) has developed the two dimensional monolithic silicon detector array named MagicPlate-512 (M512) for dosimetry of small radiation fields with spatial resolution 2 mm, an improved spatial resolution relative to commercial diode arrays. Although, the dosimetric characterisation of M512 in term of percentage depth dose and dose per pulse dependence has been completed and shows that the M512 is suitable for use as a quality assurance detector in small radiation fields [10], the angular response of the M512 detector for dose verification in arc therapy delivery has to be studied. The purpose of this study is to investigate the angular response of the M512 detector to 6 MV photon beam irradiation for different field sizes.

2. Materials and Methods

2.1. M512 and Data Acquisition system

The M512 is a two-dimensional monolithic detector array comprised of 512 active square pixels each with size 0.5 x 0.5 mm^2 and 2 mm pixel pitch. The detector is fabricated on a bulk p-type silicon substrate with thickness 470 µm and an active area of 52 x 52 mm^2. The M512 is wire bonded to a 500 µm thick printed circuit board (PCB) that provides the connection for the pixels to the readout electronics. To avoid mechanical damage, the detector array has been covered with a thin protective epoxy layer. The M512 was placed on a 5 mm thick PMMA slab and covered with a PMMA top providing a 1 mm air gap (figure 1) to maximize the detector response relative to radiochromic film for small normally incident radiation fields similar to the "air diode" [11].

The charge generated by the radiation beams in the M512 pixels was read out with a data acquisition (DAQ) system based on a multichannel (64 channel) electrometer chip, AFE0064 (Texas Instrument). The system provides accurate measurements of charge because each current integrator is equipped with a double sampler for the baseline [12]. The AFE chip enable simultaneous measurement of 512 channels and are connected to a series of analogue to digital converters (ADC), synchronized with the linac electron gun pulses and read out using a field programmable gate array (FPGA). The data is transferred from the DAQ to an in-house software interface named Magic Suite on a host computer via USB2.0 communication protocol.

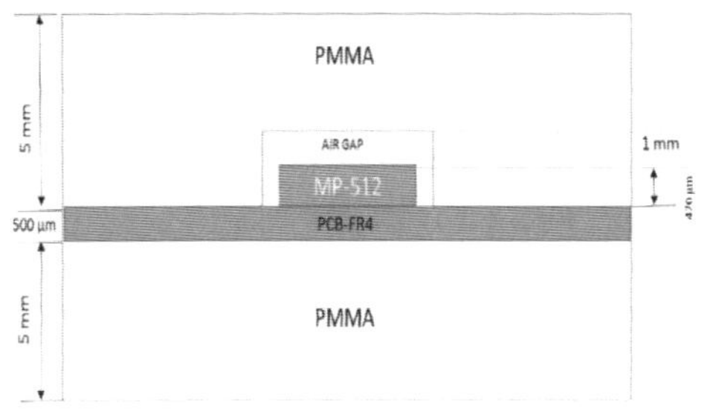

Figure 1. The cross section schematic diagram of the M512 packaging (not to scale).

Figure 2. The Experimental setup for measurement of the angular response of M512 inside the cylindrical PMMA phantom.

2.2. Angular dependence

The M512 array was placed at the iso-centre of a cylindrical PMMA phantom with a diameter of 30 cm. The detector was set in a vertical position to avoid attenuation from the couch (figure 2) and the front

IOP Conf. Series: Journal of Physics: Conf. Series **777** (2017) 012020 doi:10.1088/1742-6596/777/1/012020

surface of the detector was aligned perpendicular to the iso-centre of the radiation beam at the gantry angle 270° (incident beam angle zero). The angular response measurements were performed for square fields of sizes 3x3 cm^2 and 10x10 cm^2 using a 6 MV photon beam. A fixed dose of 100 monitor units (MU) at a dose rate of 600 MU/min was delivered for each incident beam angle from 0° to 180° in 15° increments. The angular response of M512 was determined to be the ratio over an average of the responses of 4 central pixels at a certain angle, normalized to the same pixels' response at the incident angle zero. The directional response and beam profiles acquired by the M512 were compared with EBT3 films which were irradiated at the same geometric orientation for angles 0°, 45°, 90°, 135° and 180°.

3. Results

The full-width at half maximum (FWHM) of the beam profiles between EBT3 and M512 at incident angle zero are in good agreement, within ±2%, and the penumbral width agrees within less than 1 mm (not presented in this paper)). Figure 3 shows the significant difference in angular response at the central pixels for M512 compared to EBT3 film. The maximum difference was found at the incident angle 90° to be -20% and -18.58% for a field size of 3 x 3 cm^2 and 10 x 10 cm^2, respectively. In addition, the relative response of M512 which is normalized at the incidence angle zero showed less angular dependence for the angles between 0°- 45° but showed a response change of -19% for the 3 x 3 cm^2 field size and -17.70% for 10 x 10 cm^2 field at incident angle 90^0-105^0. However, the results showed no significant difference between the reference field size (10 x 10 cm^2) and a field size of 3 x 3 cm^2 ($p>0.05$) (Figure 4).

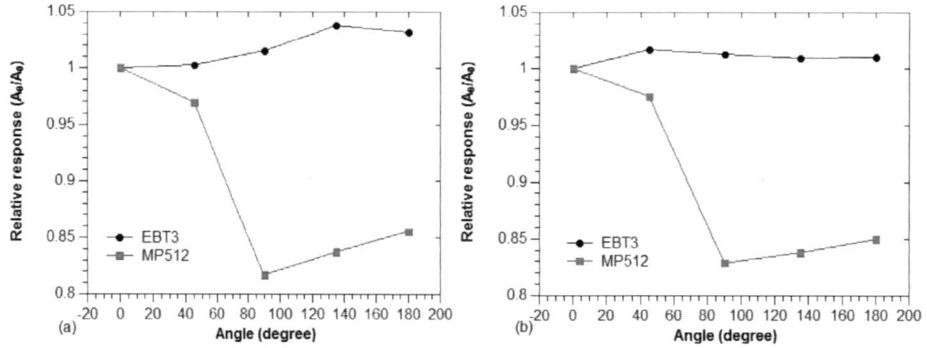

Figure 3. The comparison of the central pixel response between M512 and EBT3 for 6 MV (a) Field size 3x3 cm^2, (b) Field size 10x10 cm^2.

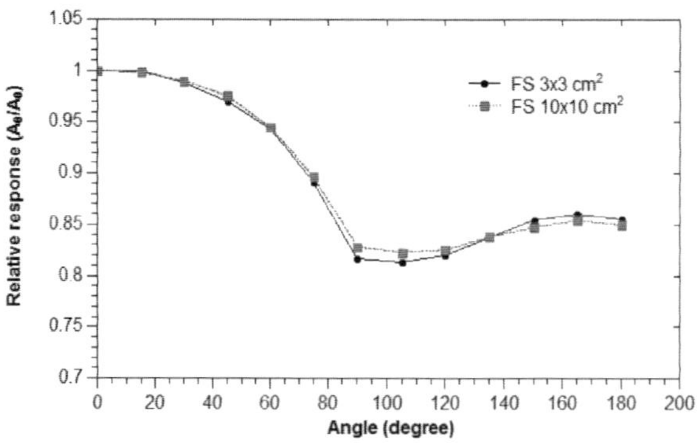

Figure 4. The angular response of a field size of 3x3 cm^2 and a field size of 10x10 cm^2.

4. Discussion and Conclusion

Although verification at gantry 0° tends to be appropriate for IMRT techniques [13], verification of treatments using complicated EBRT techniques such as VMAT and SBRT, which modulate small beamlets and gantry speed during irradiation around the patient, should be conducted considering all conditions including the beams angle of incidence. The detection of a tiny error in MLC leaf position and speed which may be transferred from the treatment planning system needs to be verified for each gantry angle in real time or retrospectively. The angular dependence of a detector must be of concern to ensure the accuracy of quality assurance. This study set out with the aim of assessing the angular dependence of a two dimensional monolithic detector array M512 and the effect of field size on its directional response.

The M512 detectors angular dependence increased as the magnitude of the incident angle increased, but showed a maximum dependence at an incident beam angle of 90°. These results are in agreement with earlier studies [7-9]. A possible reason to explain this phenomenon is the anisotropic configuration of the M512 packaging and the large size of the silicon crystal. The different material interfaces and shapes cause a variable scattering of electrons and photons into the active volume of the detector [14, 15]. The interesting result of this study found that the angular dependence of M512 showed no significant difference between a small field size of 3x3 cm^2 and the reference field size of 10x10 cm^2. However, this result has been determined through the investigation of only these two field sizes. Further field sizes should be investigated. The present results are significant to apply to the M512 for patient verification in arc therapies.

The angular correction factor is necessary to apply when using the M512 as a quality assurance tool that is easy to do as an inclinometer is providing the gantry angle continuously. Further work will be directed to clinical implementation of M512 in arc delivery verification.

5. Acknowledgement

N Stansook was supported by Ramathibodi Hospital, Mahidol University, Thailand. N Stansook would like to thank the Illawara Cancer Care Center (ICCC) staff for the useful discussion and access to the LINAC. The authors would like to thank UOW CMRP staff and students for their technical and software support.

6. Reference

[1] Das I J *et al* 2008 *Med. Phys.* **35** 206
[2] Wuerfel J U 2013 *Medical Physics International* **1** 81-90
[3] Hill R F *et al* 2008 *Rad. Meas.* **43** 1258-64
[4] Greer P B *et al* 2007 *Med. Phys.* **34** 4389-98
[5] Hill R *et al* 2010 *Med. Phys.* **37** 4355-63
[6] Wol T 1998 *Brit. J. Rad.* **71** 1168-77
[7] Jursinic P A 2009 *Med. Phys.* **36** pp. 2165-71
[8] Li Q L *et al* 2010 *Chinese Journal of Cancer* **29** 617-20
[9] Jin H *et al* 2014 *J. Appl. Clin. Med. Phys.* **15** 80-92
[10] Aldosari A H *et al* 2014 *Med. Phys.* **41** 091707
[11] Charles P H *et al* 2012 *Phys. Med. Biol.* **57** 6947-60
[12] Min S *et al* 2009 (http://www.ti.com/lit/ds/slas672/slas672.pdf)
[13] Stasi M *et al* 2012 *Med. Phys.* **39** 7626-35
[14] Werner B L *et al* 1987 *Med. Phys.* **14** 585-95
[15] Das I J and Kahn F M 1989 *Med. Phys.* **16** 367-75

A dedicated software application for treatment verification with off-line PET/CT imaging at the Heidelberg Ion Beam Therapy Center

W Chen[2], J Bauer[1,2], C Kurz[1-3], T Tessonnier[1,2], J Handrack[1,2], T Haberer[1,2], J Debus[1,2], K Parodi[1,3]
[1]Heidelberg Ion Beam Therapy Center (HIT), Im Neuenheimer Feld 450, Heidelberg 69221, Germany
[2]Department of Radiation Oncology, Heidelberg University Hospital, Im Neuenheimer Feld 400, Heidelberg 69221, Germany
[3]Ludwig-Maximilians-Universität München (LMU Munich), Department of Experimental Physics - Medical Physics, Am Coulombwall 1 (R127), Garching b. München, 85748, Germany

E-mail: wenjing.chen@med.uni-heidelberg.de

Abstract. We present the workflow of the offline-PET based range verification method used at the Heidelberg Ion Beam Therapy Center, detailing the functionalities of an in-house developed software application, SimInterface14, with which range analysis is performed. Moreover, we introduce the design of a decision support system assessing uncertainties and facilitating physicians in decisions making for plan adaptation.

1. Introduction

Due to the favourable physical properties of charged particles, ion-beam therapy is able to achieve highly conformal energy deposition in the target volume while better sparing surrounding tissue. Yet the superior conformity requires high precision of treatment delivery. Various in-vivo range verification methods have been proposed to monitor dose delivery in particle therapy [1]. Among these, positron emission tomography (PET)-based verification is being investigated clinically at the Heidelberg Ion Beam Therapy Center (HIT) [2]. The evaluation is performed by an in-house developed MeVisLab software application SimInterface14 [4][5]. This application provides various functionalities, such as data management, registration, visualization and range analysis, as well as interfaces with a remote computational cluster and a PACS system. This report gives a brief overview of the workflow of the PET range verification procedure and the functions of SimInterface14. Moreover, the structure of a decision support system is described. The purpose of this system is to assess different sources of uncertainties occurred during treatment, which can be detected in PET/CT images, and thus to assist the physicians in the decision whether plan adaptation is needed.

2. Materials and Methods

At HIT patient plans with scanned proton and carbon ion beams are optimized by the SIEMENS SyngoPT treatment planning system. Selected patients undergo PET imaging with a SIEMENS Biograph mCT full-ring PET/CT scanner for 30 minutes shortly after their treatment fraction. As the measured spatial activity distribution is not directly proportional to the dose distribution, the evaluation is based on a comparison between the measured and predicted activity distribution. The prediction is obtained from a detailed Monte-Carlo simulation and wash-out model [3]. For each patient under investigation, Monte-Carlo simulation of the expected activity distribution is performed on the treatment plan via the interface of SimInterface14 to the remote computational cluster, using the automated framework based on the Monte-Carlo code FLUKA described in [2]. The resulting isotope distributions are loaded into this MeVisLab application and a wash-out model is applied. This tissue-dependent model takes patient transferring time (from the treatment room to the PET/CT room) and data acquisition period into consideration [3]. The decayed signals from different irradiation-induced isotopes are summed up to obtain the simulated PET image. A Gaussian filter is applied to smooth the image and to mimic the finite resolution of the PET scanner.

The comparison between the acquired PET and simulated PET is carried out with SimInterface14. There are two modes for analysis – single spot and beam's eye view (BEV) evaluation. In single spot evaluation, profiles are extracted from both acquired and simulated PET images along the beam direction at a given spot within the BEV. The distal falloffs of the PET signal profiles are identified, and range shifts are estimated with the forward-zero-padding (FZP) method [5]. Furthermore, as the alignment of the acquired and simulated PET images are based on registration between treatment planning CT and PET/CT, the precision of registration affects range shift estimation. To address registration uncertainties, profiles are also extracted from the planning CT and PET/CT at the same spot and shift estimation is conducted. An example with a head patient is shown in Figure 1. In BEV evaluation, firstly a region of interest (RoI) is defined based on the planning target volume (PTV) projection to the BEV plane with user-set dilation or erosion modification of the boundaries. Then single spot evaluation is performed for each spot within the RoI. The resulting shift estimation is presented in a 2D map, as shown in Figure 2. Statistic information, such as mean, variance and standard deviation, etc. of CT and PET images is gathered for further analysis.

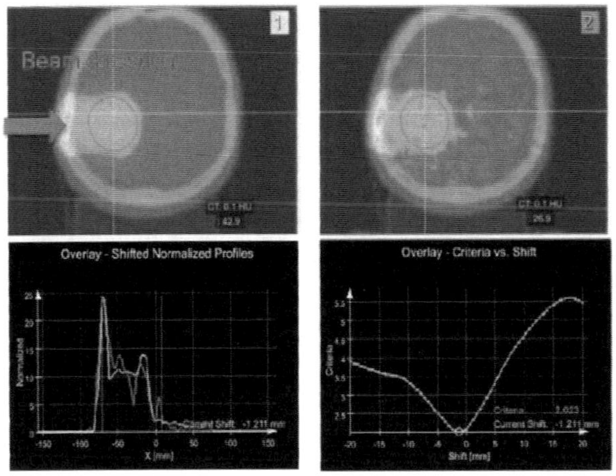

Figure 1. Top panel - A head patient's PET/CT image overlaid with measured PET (right) and the treatment planning CT overlaid with simulated PET (left) image; Bottom panel - Shift estimation for a single beam direction in beam's eye view (BEV). The profiles are displayed overlaying each other (left). Sum of absolute differences (SAD) are calculated at each shift step, resulting in a criteria curve (right). The estimated shift is the one which gives the smallest SAD in the curve.

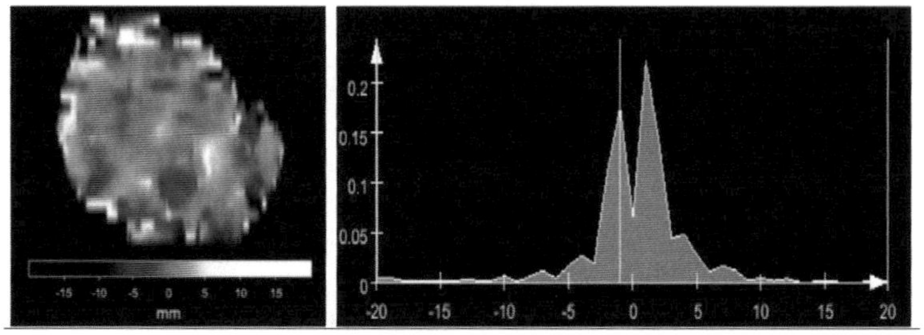

Figure 2. Estimated range shifts BEV map (left) and histogram (right) for the clinical example shown in Figure 1.

3. Results

More than 2000 patients have been treated up to now at HIT. Among them, over 200 patients have been imaged with PET after their treatments for monitoring. The indications of these patients represent the full therapeutic spectrum of HIT. Systematic analysis has been performed for 20 glioma patients, and the mean range shifts were between 2 mm and 4 mm [6]. Moreover, considering range verification for moving target, a feasibility study of time-resolved (4D) PET/CT has been carried out with liver patients. The results showed improvements in range analysis with 4D-PET/CT for targets with superior-inferior motions larger than 10 mm [7]. Evaluation of range uncertainties of the remaining cases is currently on-going.

In addition to patient studies with the already implemented functionalities, the development of SimInterface14 continues. The final goal is to develop a decision support system (DSS) which gives suggestions if plan adaptation is needed. Figure 3 shows a schematic drawing of the designed structure of the DSS. The inputs of the system are 2D maps of range shift estimations and statistics information within the RoI. The maps can be generally categorized into 3 types: maps to estimate the range shifts, maps to evaluate the reliability of Monte-Carlo simulated PET, and maps to determine the reliability of the estimated shift. For each map, a threshold is set according to the parameter represented by the map. If the value of a voxel is within the given range, this means that the voxel passes certain criteria and it is assigned with value 1. Otherwise the failed voxels are set to 0. The resulted binary maps are multiplied to each other and the "shift and reliability map" (SRM) is created. Based on the information of SRM, the system can provide insights on the necessity of plan adaptation or replanning. To trace back the uncertainties occurred during the treatment and shift estimation, one only needs to look for the 0 values in the SRM and trace back where the 0 values came from. In this way, large shifts of beam range (true positive) or unreliability in PET simulation and/or shift estimation (false positive) can be likely disentangled.

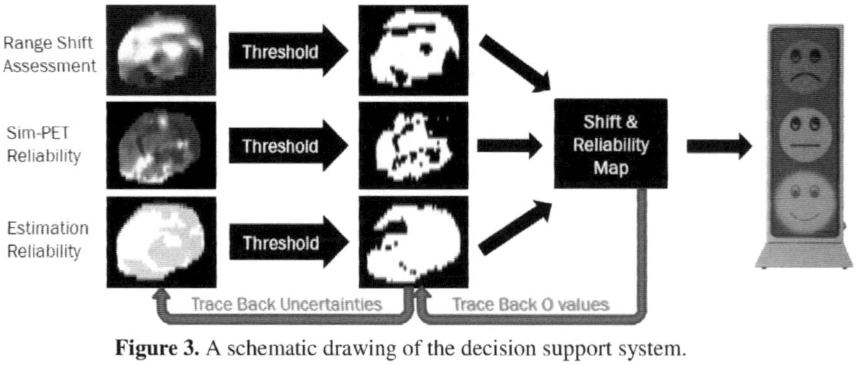

Figure 3. A schematic drawing of the decision support system.

4. Conclusions

We present an overview of the implementation and recent progresses of our PET/CT-based range verification application in a clinical setting. The current range verification method and the functionalities of SimInterface14 have been described. Furthermore, we introduce the idea of a decision support system, which is able to interpret the results of PET/CT-based verification and track various sources of uncertainties. The ultimate objective is to provide a reliable statement on the measured beam range, and thus support decision making processes towards adaptive particle therapy.

5. Acknowledgements

This work is supported by BMBF (SPARTA, grant agreement number 01IB13001G) and DFG (KFO214). We acknowledge A. Köhn, F. Weiler, C. Brachmann, D. Corr, JM Kuhnigk, S. Wirtz, S. Krass from FME, D. Unholtz, K Frey formerly from HIT, L. Zhang and C. Hentschke from DKFZ.

6. References

[1] Knopf A and Lomax A 2013 Phys. Med. Biol. **58** R131-60
[2] Bauer J *et al* 2013 Radioth. *Oncol.* **107** 218-26
[3] Parodi K *et al* 2007, *Int, J. Rad. Oncol. Biol. Phys.* **68** 920-34
[4] Unholtz D *et al* 2011 *Conference Record of IEEE* NSS-MIC
[5] Chen W *et al* 2015 PTCOG **54** poster
[6] Nischwitz S *et al* 2015 *Radiother. Oncol.* **115** 179-85
[7] Kurz C *et al* 2016 *Med. Phys.* **43** 975

Optimisation of output factor measurements using the Magic Plate 512 silicon dosimeter array in small megavoltage photon fields

K Utitsarn[1], Z A Alrowaili[1], N Stansook[1], M Lerch[1], M Petasecca[1], M Carolan[2], A Rosenfeld[1]
[1]Centre for Medical Radiation Physics, University of Wollongong, Australia
[2]Illawarra Cancer Care Centre, Wollongong hospital, Australia

E-mail: ku566@uowmail.edu.au

Abstract. We evaluate the impact of an air gap and optimization of this air gap for the MP512 silicon detector array when operated in dosimetry mode for small photon field measurements in solid water. We present output factor measurements for 6MV and 10 MV photon beams with the square field sizes ranging from 0.5 to 10 cm^2. The size of the air gap above the MP512 detector was changed from 0.5, 1.0, 1.2, 2.0 and 2.6 mm. We compare the output factors measurements of the MP512 with EBT3 film and the MO*Skin* dosimeter. For the two photon energies investigated, we find that the output factor measured by the MP512 reduce with increasing air gap and reducing of field size. The reduction in output factor is most pronounced for the 0.5 and 1 cm^2 field sizes. The air gap of 0.5 mm and 1.2 mm showed good agreement with the EBT3 film and MO*Skin* output factor for 6 and 10 MV photon fields, respectively. The negligible effect on dosimetry for the field sizes larger than 4x4 cm^2 demonstrates that the electronic disequilibrium caused by small air gap only influences the dosimetry measurements for small fields. The study shows that the output factor reduction is enhanced by increasing of air gap and demonstrates that the optimal air gap for the MP512 at 6 and 10 MV photon fields is 0.5mm.

1. Introduction

Modern radiation treatment techniques such as Stereotactic Radiosurgery (SRS) and Stereotactic Body Radiation Therapy (SBRT) are increasingly being used in cancer treatment. Large doses per fraction of SRS/SBRT are delivered in small photon radiation fields leading to steep dose gradients, Quality assurance (QA) is needed to ensure that the delivered dose matches the plan dose distribution [1]. Current commercial dosimetric QA tools for SRS/SBRT include ionization chambers, used for point dose measurements, as well as two-dimensional diode arrays and pixelated ionization chamber arrays. One limitation of the ionization chamber is that the dosimetry in small field can be perturbed by the chamber volume averaging effect [2]. Film has good spatial resolution but their processing is time consuming. ArcCHECK (Sun Nuclear, Melbourne, FL) contain an array of 1386 diodes distributed in a cylindrical acrylic phantom. To perform the patient QA, the treatment plan needs to be transferred to the QA system. The Delta4 (ScandiDos, Uppsala, Sweden), which consist of 1069 silicon diodes in a crossed array inside a cylindrical polymethylmethacrylate (PMMA) phantom, requires the recalculation of the dose on a CT scan phantom for the verification procedure. The operation system of both ArcCHECK and Delta4 is performed on the treatment couch [3] and is not designed to operate in transmission mode. In small irradiation field dose delivery via SRS and SBRT, the QA procedure needs to have a detector with high spatial resolution. The Magic Plate 512 (MP512) a two dimensional monolithic silicon detectors array has a spatial resolution of 2 mm (pitch of array elements), which is suitable for such small field dosimetry. The patient treatment verification can be performed without the need to transfer the treatment plan to the detector and real time dose measurement can be made [4]. The purpose of this study is to examine the effect a small air gap immediately above surface of the MP512 has on the measured output factors for 6 and 10 MV photon fields as part of QA in SBRT and SRS.

2. Methodology

MP512 is a silicon monolithic detector manufactured on a p-type substrate. The MP512 array consists of 512 pixels with size 0.5 x 0.5 mm^2 and pitch 2 mm with an overall dimension of 52 x 52 mm^2. The MP512 monolithic detector is wire bonded on a printed circuit board 0.5 mm thick and covered by a thin layer of raisin to protect the bonding wires as well as the silicon detector surface from moisture [4]. The output factor measured with MP512 is compared with similar measurement using EBT3 films as well as MO*Skin* dosimeters. The MP512 was placed on a solid water phantom at the depth of 10 cm with 10 cm additional back scatter material and was aligned at the centre of the beam. The measurements were carried out with size of the air gap above the detector set at 0.5, 1.0, 1.2, 2.0 and 2.6 mm. The field size was varied from 0.5 x 0.5 cm^2 to 10 x10 cm^2 while SSD was 90 cm. The measurements were performed on a Varian 21XI linear accelerator with 6 MV and 10 MV photon beams and dose rate 600 MU/min. In each case 100 MU was delivered with open field.

3. Results

Figure 1 shows the output factor for 6 MV measured by MP512 at 10 cm depth in a PMMA phantom for different air gaps above detector. Also shown air data from equivalent measurements performed using EBT3 film and MO*Skin*, with no air gap above them however. All measurements air normalized to the response in a 10 x 10 cm^2 field size. For small field sizes (0.5 and 1 cm^2) at 6 MV, the MP512 measured output factor reduces with increasing air gap above the detector. This reduction is significant when the air gap increases beyond 2mm. For larger field sizes ($\geq$2x2 cm^2) at 6 MV, we observe a small but increasing effect on the measured output factor with increasing air gap. The optimum MP512 air gap is 0.5 mm as it shows good agreement with the EBT3 film and MO*Skin* output factor. Figure 2 shows the output factor for 10 MV photons measured by MP512 with different air gaps above detector compared to the EBT3 film and MO*Skin* (with no air gap), normalised to the MP512 response at 10 x10 cm^2 field size. For all field sizes at 10 MV, the MP512 measured output factor reduces with increasing air gap above the detector. The output factor measured by MP512 with air gap size of 1.2 mm seems to matches best to the output factors measured with the EBT3 and MO*Skin*.

4. Conclusion

The output factors measured by the 2D detector array MP512 with different air gaps immediately above the detector placed at a depth of 10 cm in solid water phantom have been measured in both 6 and 10 MV photon fields. The properties of the array confirmed suitability for use as a phantom dosimeter for QA in SRS and SBRT. The output factor measured by MP512 with an air gap of 0.5 mm and 1.2 mm matches to within 2% that measured by GAF™ film and MO*Skin*™ for field sizes down to 0.5x0.5 cm^2, for 6 and 10 MV, photon field respectively.

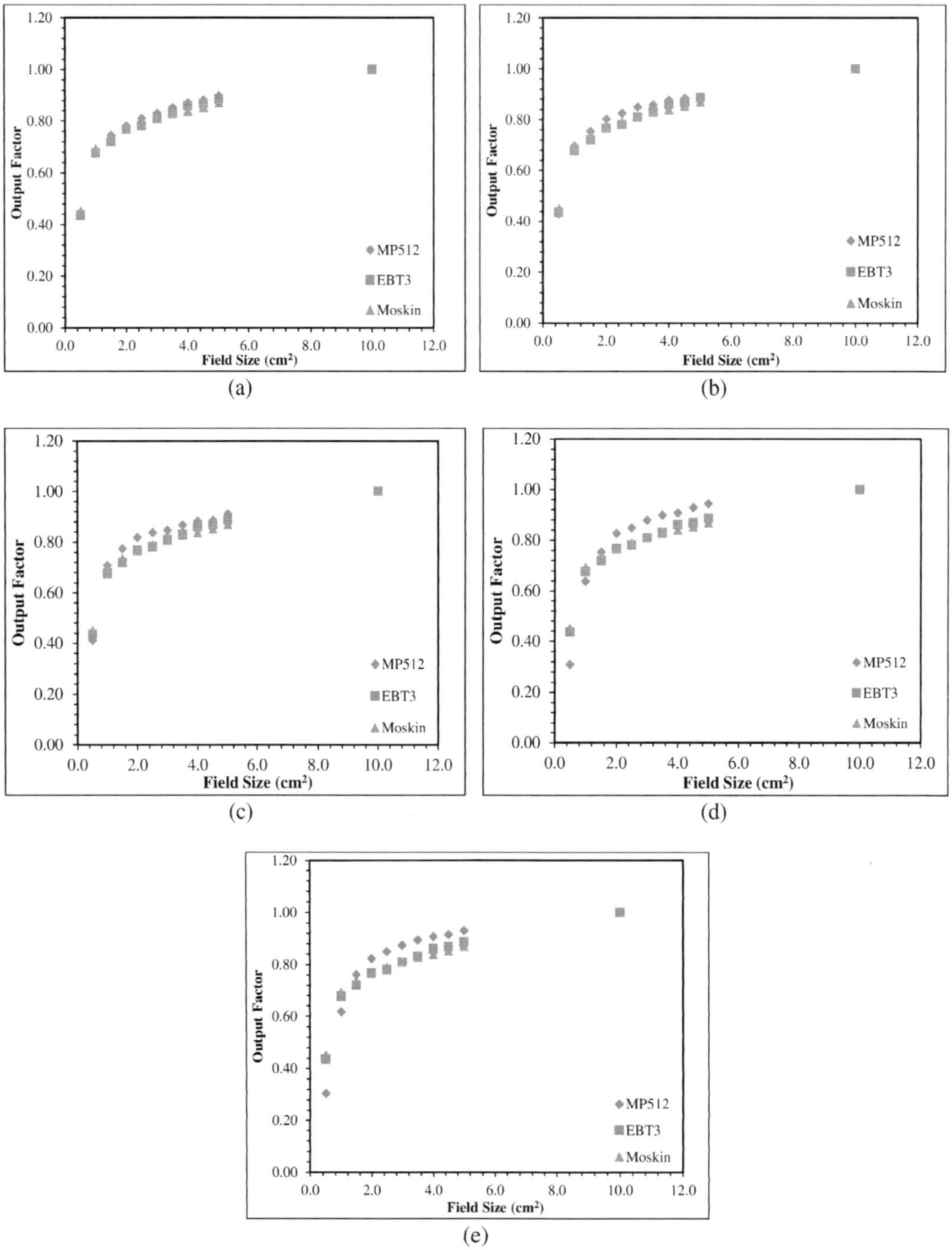

Figure 1. Field size response of MP512, EBT3 film and MO*Skin* for 6 MV photon beam normalized to the response at 10 x 10 cm^2 field size at depth 10cm in a PMMA phantom for different airgaps. (a) 0.5 mm, (b) 1.0 mm, (c) 1.2 mm, (d) 2.0 mm and (e) 2.6 mm

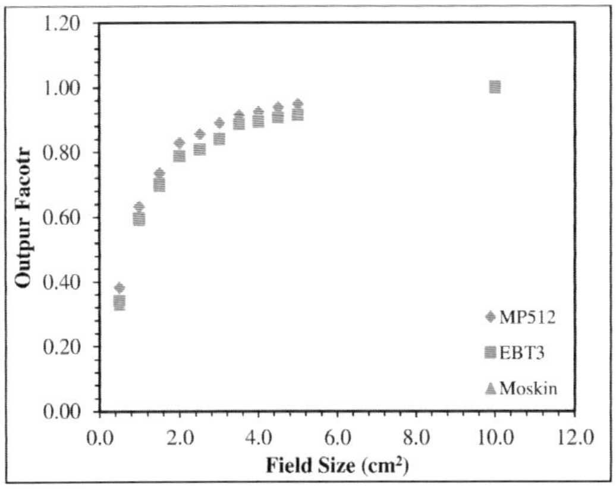

(a)

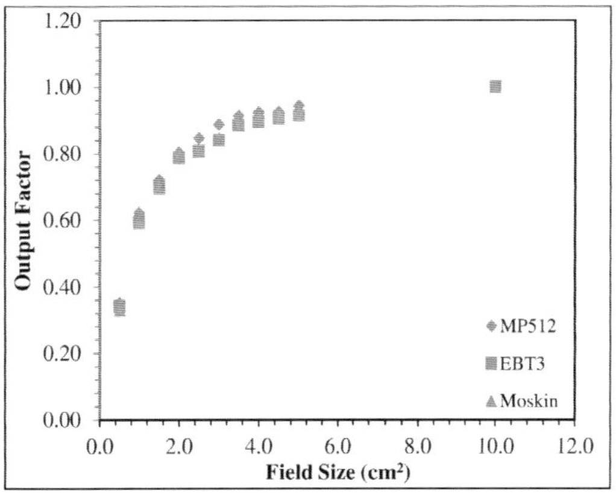

(b)

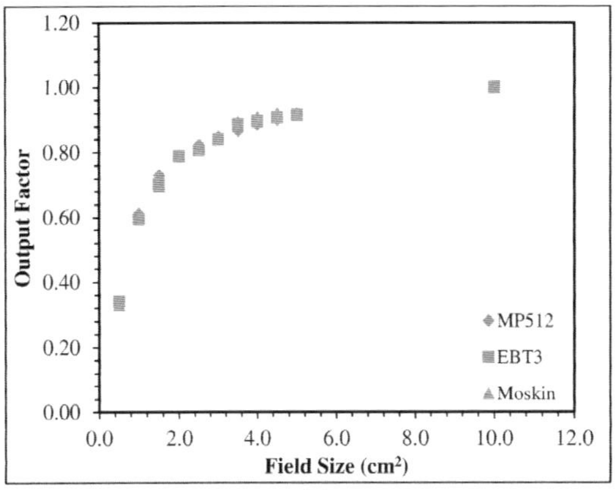

(c)

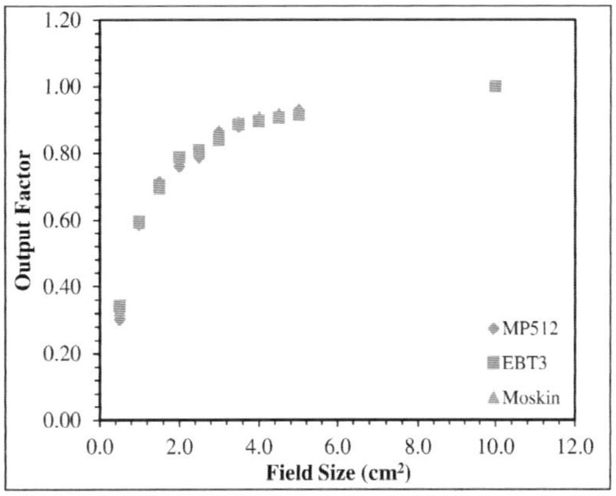

(d)

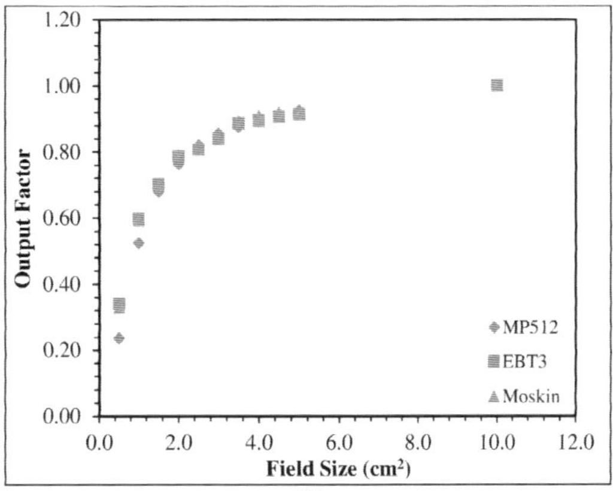

(d)

Figure 2. Field size of MP512, EBT3 film and MO*Skin* for 10 MV photon beam normalized to the response at 10 x 10 cm^2 field size at 10 cm depth in PMMA phantom. (a) 0.5 mm, (b) 1.0 mm, (c) 1.2 mm, (d) 2.0 mm and (e) 2.6 mm

5. Reference

[1] Martin A and Gaya A 2010 *Clin. Oncol.* **22** 157-72
[2] Low D A *et al* 2003 *Med. Phys.* **30** 1706-11
[3] Li G *et al* 2013 *J. Appl. Clin. Med. Phys.* **14** 13-24
[4] Aldosari A H *et al* 2014 *Med. Phys.* **41** 091707

OpenPET: a novel open-type PET system for 3D dose verification in particle therapy

T Yamaya
[1] National Institute of Radiological Sciences, National Institutes for Quantum and Radiological Science and Technology, 4-9-1 Anagawa, Inage-ku, Chiba 263-8555, Japan

Email: yamaya.taiga@qst.go.jp

Abstract. The OpenPET is the world's first open-type 3D PET scanner for PET image-guided particle therapy such as *in situ* dose verification and direct tumour tracking. Even with a full-ring geometry, the OpenPET has an open gap between its two detector rings through which the treatment beam passes. Following the initial proposal of the dual-ring OpenPET (DROP), the single-ring OpenPET (SROP) was also proposed as a more efficient geometry than DROP in terms of manufacturing cost and sensitivity. A small SROP prototype was developed and feasibility of visualizing a 3D distribution of beam stopping positions inside a phantom was shown with the help of radioisotope particle beams, used as primary beams. Following these results, a full-size whole-body SROP prototype was developed.

1. Introduction

Cancer is a major cause of death in developed nations, and the disease death rate continues to increase. Therefore, many efforts have been made toward better diagnosis and better treatment. Among diagnostic methods, positron emission tomography (PET), which can visualize functions such as metabolism by injecting radioisotope tracers labelled with positron emitters, is expected to enable earlier and more precise cancer diagnosis.

For treatment, on the other hand, radiotherapy is essential for effective cancer treatment with minimized side effects. Specifically, particle therapy such as proton and carbon ion therapy are expected to be the ultimate radiotherapy because they can concentrate the dose even in a deep tumour. Thus there has been remarkable progress in PET and radiotherapy, but no one has looked into the great potential to be obtained by the combination of both.

An open-type PET geometry (OpenPET) is a new idea to visualize a physically opened space between two detector rings [1], which will move researchers toward a future joint PET imaging and radiotherapy system. OpenPET is expected to achieve in-beam PET, which is a method to monitor *in situ* charged particle therapy. Without injecting any PET tracer, positron emitters are produced through fragmentation reactions between the projectiles and the atomic nuclei of the tissue during patient irradiation [2-3]. Compared with conventional radiation therapy, charged particle therapy can highly concentrate the dose in a tumour. This means if there is any difference between the actual irradiation and the treatment plan, the tumour treatment will be compromised, and the normal tissue around the tumour will be damaged. Therefore, quality of treatment must be assured by in-beam PET.

2. Simulation and system design

We have proposed two geometries for OpenPET: dual-ring OpenPET (DROP) [1] and single-ring OpenPET (SROP) [4]. Figure 1 shows a conceptual illustration of the SROP comparing with the DROP and a conventional PET which is positioned at a slant angle relative to the bed to form an accessible space (hereinafter referred to as "slant PET") [5]. Compared with the slant PET, the SROP is expected to provide higher sensitivity with a smaller number of detectors. Thus, in the case of in-beam PET, provides a closer positioning of the beam port, which minimizes beam broadening.

We calculated the geometrical sensitivity at the centre. As fixed parameters, the bed width of 600 mm and open gap of 300 mm were selected to provide a sufficient open space during ion-beam therapy. Here, we assumed that the area of the block-detector was 2500 mm^2 (i.e., 50 mm × 50 mm). When the detector surface area was 500,000 mm^2, in which the equivalent number of block-detectors was 200, the geometrical sensitivities of the SROP, DROP and slant PET were 27.8%, 23.9% and 22.1%, respectively. The SROP can provide an accessible and observable open space with higher sensitivity and a reduced number of detectors compared to the previous generation geometries.

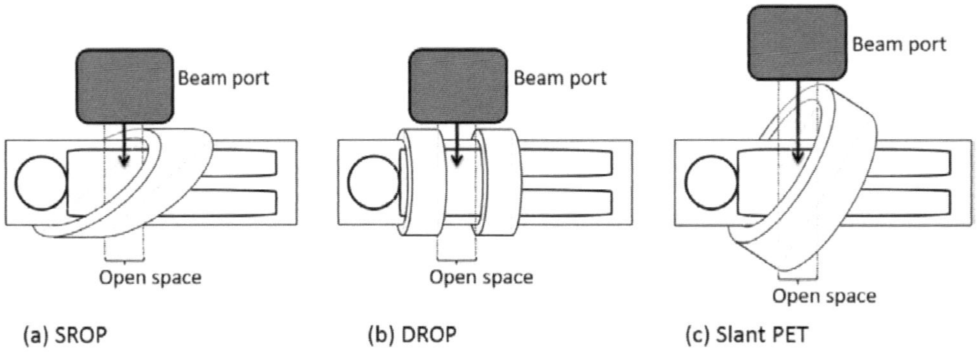

Figure 1. Conceptual illustrations of PET geometries with an accessible open space to the patient.

Figure 2. Photographs of the small SROP prototype taken in the closed conventional PET (a) and OpenPET (b) modes. The transformation was completed within 10s.

3. Development of a small prototype

Figure 2 shows a small SROP prototype [6], which had two open and closed modes. The open mode formed the SROP geometry and the closed mode formed the conventional cylindrical PET geometry. Sixteen detector units each of which consisted of two depth-of-interaction (DOI) detectors were arranged to form a perfect circle with a diameter of 250 mm. Each DOI detector consisted of a H8500 PMT and the 4-layer 16 × 16 array of Zr-doped GSO (GSOZ) scintillators with a size of 2.8 × 2.8 × 7.5 mm^3. Detector units had an axial-shift mechanism so that they could be transformed into the SROP having an open space of 139 mm. In the open mode, the centre of each detector surface, positioned on the parallel planes, was slanted 45° against the axial direction. For shifting purposes, each detector unit was connected to the neighbouring detector units by linear guides. Transformation between the cylindrical PET and the SROP, which was controlled by one rotation handle was completed within 10s.

We evaluated the basic performance of the prototype. The spatial resolution and sensitivity were 2.6 mm and 5.1% for the open mode and 2.1 mm and 7.3% for the closed mode. The SROP enables in-beam PET imaging at a slight cost of imaging performance. The decrease of the performance can be minimized, for example, by transforming into the close mode immediately after the irradiation, while maintaining the open space only for the in-beam PET measurement.

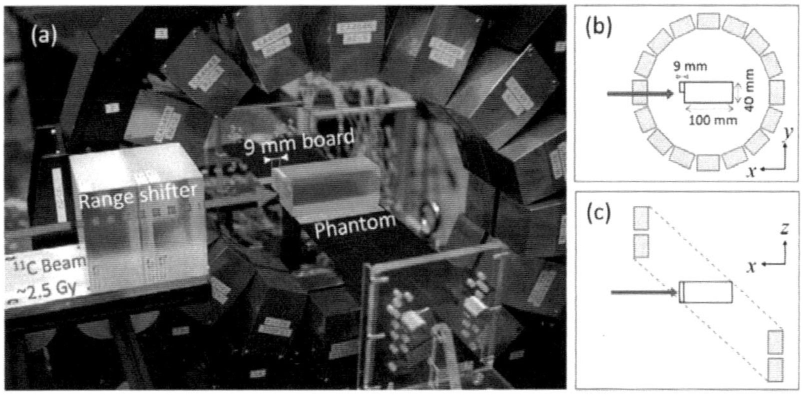

Figure 3. Setup for PMMA phantom irradiation by the ^{11}C beam and in-beam PET measurements. A photograph of the setup (a) and illustrations for front (b) and top (c) views are shown.

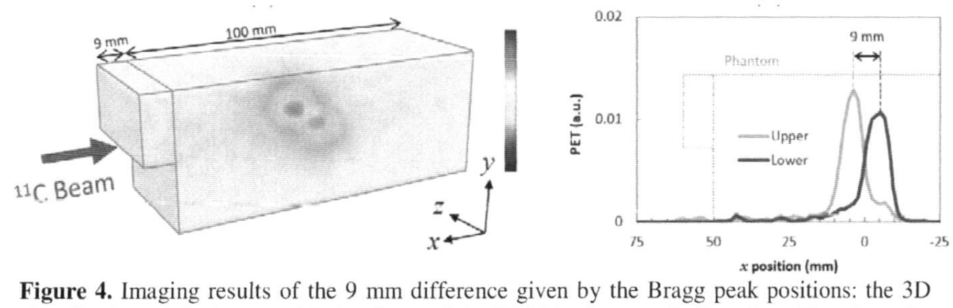

Figure 4. Imaging results of the 9 mm difference given by the Bragg peak positions: the 3D visualization of the image with the shape of the PMMA phantom (left) and profiles along the beam direction (right).

4. In-beam imaging of carbon ion beam

The small prototype was brought into the heavy ion medical accelerator in Chiba (HIMAC), where we conducted the in-beam imaging tests. A PMMA phantom was irradiated with an RI beam of ^{11}C, which has a half-live of about 20 min. The RI beam was generated as a secondary beam from the ^{12}C irradiation onto a Be target [7]. Figure 3 shows the experiment setup. The dimensions of the rectangular cuboid PMMA phantom were 40 × 40 × 100 mm^3 and an additional PMMA board with a thickness of 9 mm was used to cover the upper half of the beam. The irradiation time was about 10s and a dose of about 2.5 Gy was given. The beam intensity was about 5×10^6 particles per second and the beam energy was 340 MeV/u. Therefore, the Bragg peak position in PMMA was at 168 mm. A PMMA range shifter of 115 mm thickness was used so that the Bragg peak position was moved to 53 mm from the entrance of the phantom. PET data were measured during the irradiation and for 20 minutes after the irradiation ended. Because the irradiation beam in the HIMAC had a cycle of 3.3 s with beam-on time (spill-on) of about 1.8 s and a beam-off time (spill-off) of about 1.5 s, PET data for the spill-off time were extracted. The 3D list-mode maximum likelihood expectation maximization (LM-MLEM) was employed for the PET image reconstruction. The number of iterations was 50 and the voxel size was $1.5 \times 1.5 \times 1.5$ mm^3. Random correction and attenuation correction were applied but scatter correction was not applied because the phantom size was small enough so that the scatter effects can be ignored.

Figures 4 shows images reconstructed from counts measured between spills and for 20 minutes after the irradiation. The number of counts was 626 k counts. Figure 4 also shows the profiles of single voxel lines along the beam direction for the upper part with the 9 mm PMMA board and for the lower part without it. The peak position of the lower part was calculated by parabola fitting as 55.0 mm (planned Bragg peak at 53 mm). The difference in peak positions was clearly observed and measured as 8.8 mm, which was almost the same as the given difference. Following proof-of-concept results, a full-size whole-body SROP prototype was developed (figure 5).

Figure 5. The full-size whole-body SROP prototype.

5. Conclusion

This paper described basic concepts and prototypes of the OpenPET, which led to the world's first open-type 3D PET scanner. The SROP is a more efficient OpenPET geometry in terms of the gap size and the sensitivity. Demonstrations with a prototype showed a proof-of-concept of PET-image guided particle therapy such as *in situ* dose verification.

6. References

[1] Yamaya T *et al* 2008 *Phys. Med. Biol.* **53** 757
[2] Enghardt W, Crespo P, Fiedler F, et al 2004*Nucl. Instrum. Methods Phys. Res. A* **525** 284
[3] Parodi K *et al* 2005 *IEEE Trans. Nucl. Sci.* **52** 778
[4] Tashima H *et al* 2012 *Phys. Med. Biol.* **57**: 4705
[5] Crespo P *et al* 2006 *Phys. Med. Biol.* **51** 2143
[6] Tashima H *et al* 2016 *Phys. Med. Biol.* **61** 1795
[7] Urakabe E *et al* 2001 *Japan. J. Appl. Phys.* **40** 2540

What does large randomized trials tell us about the fractionation sensitivity of prostate cancer?

M Høyer
Danish Center for Particle Therapy, Aarhus University Hospital, 8000 Aarhus C, Denmark

E-mail: hoyer@aarhus.rm.dk

Abstract: Seven randomized clinical trials have tested the use of moderate hypofractionation (2.1-3.5 Gy per fraction) compared to conventional fractionation (1.8-2.0 Gy per fraction) in radiotherapy for localized prostate cancer. The trails find that moderate hypofractionation results in acceptable PSA-control and morbidities that are comparable to conventional fractionation schedules. Extrapolation of the results from the earliest randomized trials indicated a low α/β-values for prostate cancer, but the more recent – and large – studies suggest that the value is moderately or considerably higher. Moderate hypofractionation schedules (ie. 20 x 3 Gy) are now being implemented in routine practice. They are cost-effective and convinient, but they are not consistantly concordant with fractionation sensitivity parameters extrapolated from clinical trials on moderate hypofractionation of prostate cancer.

1. Introduction

Standard external beam radiation therapy (EBRT) is delivered with 1.8-2.0 Gy/fractions with 5 weekly fractions over 7-9 weeks. There are several arguments for use of more hypofractionated schedules. Years ago it was lack of ressources, whereas nowadays the competition with robotic assisted prostatectomy is the driving force behind hypofractionation. However, there are also biological arguments for chosing hypofractionation in frequently occurring cancers such as breast- and prostate cancer. According to these, the therapeutic ratio improves by use of hypofractionation of localized prostate cancer due to a high fractionation sensitivity (low α/β-value) of the slowly proliferating prostate cancer cells.

Modeling of clinical studies have suggested a low α/β-values for breast and prostate cancer. Brenner and Hall suggested that the α/β-value for prostate cancer was 1.5 Gy (0.8-2.2) in a study comparing isoeffective doses from EBRT, low- (LDR) and high dose rate (HDR) brachytherapy studies [1]. The endpoint was PSA relapse and patients were stratified by pretreatment PSA level. The study was criticized for the inclusion of brachytherapy data, but Mirabell et al found a similarly low α/β-value with narrow confidence interval of 1.4 Gy (0.9-2.2) in a modeling study of EBRT alone stratified for risk groups [2].

Moderate hypofractionation most often refers to the use of doses in the range 2.1-3.5 Gy per fraction and extreme hypofractionation with dosed above 3.5 Gy per fraction. The present review focus on randomized trials where the experimental arm is moderate hypofractionation and the comparator is standard fractionation.

2. Conventional fractionated radiation therapy of prostate cancer

Randomized clinical trials have demonstarated the efficacy of radiation therapy delivered with normofractionation. In a recent update of the Scandinavian Prostate Cancer Group 7 trial, the 15-year cancer-specific mortality reduced from 34% to 17% and the overall survival was prolonged by 2.4 years when radiation therapy was given in combination with life-long bicalutamide to high risk prostate cancer compared to the antiandrogen therapy alone [3]. The improved cancer-specifik survival was found in all tumor stage-, PSA- and age-groups. The morbidity related to the treatment was manageable. There was moderate - but significantly more - urinary, bowel and sexual bother in radiation therapy group.

More detailed analysis of the morbidity based on patient reported outcome measures (PROM) following radiation therapy with conformal techniques and conventional doses for prostate cancer reveale more morbidity and bother related to urinary-, bowel- and sexual function than we measure with use of physician administrated scoring systems. In a long-term morbidity study, some patients experience incontinence for solid stool, ability to defer defecation for less than 15 minutes, unproductive call for stool, clustering of stool and mucus in stool to minor degree [4]. These five factors are furthermore related to a reduced quality of life in those who suffered. It is an important message that detailed PROMs may identify morbidity that is not revealed from the less sensitive traditional scoring systems. When new radiation technology is introduced we should be avare that it potentially may have slightly worse morbidity profile than our conventional therapy.

3. Moderate hypofractionation of prostate cancer

Seven randomized clinical studies compare moderate hypofractionation with conventional fractionated radiation therapy for prostate cancer (table 1). They vary in size, inclusion criteria and use of anti-androgen deprivation therapy (ADT) as well as in design. Importantly, most studies have different equivalent dose in 2-Gy fractions (EQD2) between the arms when assuming an α/β-value of 1.5 Gy. As an example, there is a 12 Gy difference in EQD2 between the two randomization arms of the HYPRO study.

Two studies used low radiation dose in EBRT for low- and intermediate risk prostate cancer [5;6]. The EQD2 was higher than the conventional arm in the Lukka study whereas the hypofractionated arm was the hottest in the Yeoh study. With a non-inferiority design, the Lukka study concluded that biochemical control in the hypofractionated arm was not non-inferior and might be inferior to conventional fractionation. In contrast, the hypofractionated arm had superior relapse free survival in the Yeoh study. Differences in EQD2 between randomization arms may explain the results of the two trials, and the studies are in accordance with a low α/β-value ranging 1-2 Gy for prostate cancer.

The Italian Arcangeli study compared two arms with comparable EQD2 in high-risk patients who all received 9 months ADT [7]. This study has a relatively low power (n=168) and was published with a relatively short median follow-up time of 32 months, but it found a clear improvement in biochemical progression free survival in patients treated with hypofractionation. This study therefore favours a very low α/β-value. The treatment related morbidity was modest and there was no difference between the two fractionation arms.

The Pollack study compared two schedules with a difference in EQD2 of 6 Gy where the hypofractionated arm was the hottest [8]. The study included all risk groups. The CTV-PTV margin was slightly smaller in the hypofractionated arms. They found no difference in the combined endpoint of biochemical and clinical failure between the arms assessed with both the American Society for Therapeutic Radiation Oncology (ASTRO) and the nadir+2 principles. Considering the higher EQD2 in the hypofractionated arm, this study does not favour a low α/β-value. Acute and late morbidities were acceptable and not different between the arms, but a subgroup of patients with considerable pretreatment lower urinary tract symptoms suffered more long-term side-effects when treated with hypofractionation.

The UK CHHiP non-inferiority study was presented at the ECC2015 in Vienna [9]. Dearnaley et al. compared three arms where the conventional 74 Gy arm had comparable EQD2 to the 57 Gy in 19 fractions arm with an α/β-value of 1.5 Gy and to the 60 Gy in 20 fractions arm with an α/β-value of 2.5 Gy. The study population had a <30% risk of vesicular invasion and was primarily in the intermediate-risk group. The study showed non-inferiority between the 20 fraction arm and the conventional arm and therefore in accordance with an α/β-value in the range of 2.5 Gy and not an ultra-low value. In the evaluation of the toxicity, there as no difference in the severity of acute and late side-effects with the exception of earlier appearance of acute effects in the hypofractionated compared to the conventional schedule. PROM data from the CHHiP trial with a median follow-up of more than 4 years revealed relatively mild morbidity in all three arms and no differences between the them [12].

Table 1. Randomized clinical trials comparing moderate hypofractionation to conventional fractionated radiotherapy for prostate cancer.

Author	N	RT schedule	EQD2 (Gy)	Endpoint and Efficacy	Morbidity
Lukka [5]	936	66 Gy/33 frx 52.5 Gy/20 frx	66 62	BCF 53% vs. 60% HR=1.18 (0.99-1.41) hypo not non-inferior	More acute morbidity with hypo
Yeoh [6]	217	64 Gy/32 frx 55 Gy/20 frx	64 67	BCF 34% vs. 53% Hypo best	More GU morbidity with conv
Arcangeli [7]	168	80 Gy/40 frx 62 Gy/20 frx	80 81	4-year FFBF[a] 82% vs. 60% Hypo best	NS
Pollack [8]	303	78 Gy/39 frx 70.2 Gy/26 frx	78 84	BCF 23% vs. 21%. NS	NS
Dearnaley [9]	3216	74 Gy/32 frx 57 Gy/19 frx 60 Gy/20 frx	74 74 77	5-yr PFR HR=0.83 (0.68-1.03)[b] HR=1.20 (0.99-1.45)[b] Hypo (60 Gy) not inferior	NS
Lee (RTOG) [10]	1092	74 Gy/41 frx 70 Gy/28 frx	70 80	DFS HR=0.85 (0.64-1.14) NS	More late GU and GI morbidities with hypo
Incrocci (HYPRO) [11]	800	78 Gy/39 frx 64.6 Gy/19 frx	78 90	DFS HR=0.86 (0.63-1.16) NS	NS

[a]Late morbidity was primary endpoint
[b]90% CI; other HR are given with 95% CI
Abbreviations: EQD2: equivalent dose in 2 Gy fractions under the assumption of α/β=1.5 Gy; BCF: combined biochemical and clinical failure rate; FFBF: freedom from biochemical failure; PFR: progression free rate; DFS: disease free survival; HR: hazard ratio; NS: non-significant; vs: versus.

Two large randomised, non-inferiority, phase III studies were presented at the ASTRO 2015 in San Antonio [10;11]. The RTOG 0415 study presented by Lee et al. and the Dutch HYPRO study presented by Incrocco et al. including low-risk and intermediate- to high-risk localised disease, respectively. The RTOG 0415 study arms were designed to be biologically iso-effective at the α/β-value of 10 Gy. The primary end-point, disease free survival (DFS), with a median follow-up of 5.8 years was 86% and 85%, which was not statistically different. Late genitourinary and gastrointestinal side effects were slightly more prevalent in the hypofractionated arm. With a median follow-up of 5 years, the HYPRO study primary end-point, relapse free survival, was 80% and 77% in favour of hypofractionation, but not statistically different from the conventional arm. Preliminary results from the toxicity analysis revealed no differences in early and late toxicity between the two arms. In both these studies, the EQD2 was considerably higher in the hypofractionated arm when assuming a low α/β-value of 1.5 Gy, and with non-inferiority both studies indicate a much higher α/β-value.

4. Discussion

Consistently, the seven studies showed that moderate hypofractionation lead to acceptable biochemical control, and the largest of the studies concluded that moderate hypofractionation with sufficient dose was not inferior to the conventional fractionated regiment. It was most clearly shown by the CHHiP trial where the 20 x 3 Gy regimen was non-inferior to the conventional 37 x 2 Gy regimen whereas the 19 x 3 Gy was not non-inferior to the conventional schedule. All studies showed that moderate hypofractionation with the right doses and delivered with modern IMRT and IGRT was associated with tolerable morbidity. In some of the studies, there was slightly more late morbidity in the hypofractionated arms, but the additional morbidities were in most cases grade 2 and relatively mild. Moderate hypofractionation seems to be safe and several institutions have adopted the moderately hypofractionated schedules. However, it will be important to follow the patients in long-term follow-up with use of sensitive scoring systems such as the PROM questionnaires.

Modeling studies have indicated that prostate cancer may have a very low α/β-value in the range of 1.5 Gy which favours hypofractionation. Slow proliferation may be associated to the high fractionation sensitivity and low rate of proliferation has been shown in prostate cancer [13]. The early randomized trials were in accordance with the modeling studies, all indicating a high fractionation sensitivity. In contrast, the more recent trials are inconsistant with the earlier studies. Among these are three large studies, CHHiP, RTOG0415 and HYPRO indicating moderately or considerably higher α/β-values compared to the early randomized trials and to the modeling studies. These inconsistancies are puzzleing and could be explained by differences in risk groups, Gleason grades and use of ADT between the studies. Intuitively, low α/β-value is expected in low Gleason-grade and low-risk patients in general, but there are no such correlation found between the seven randomized studies. Neither is there a correlation between fractionation sensitivity and use of ADT.

There are a number factors to disclaim when comparing studies on moderate hypofractionation. The variation between the studies in terms of design, total doses, fractionation schedules as well at the differences between the study populations may be important confounders. Furthermore, the hypofractionated schedules were delivered with shorter overall treatment times and the shortening of overall treatment time differ between the studies.

These trials open a new discussion on the α/β-value of prostate cancer and results on more extreme hypofractionation of prostate cancer are awaited. Among these is the Scandinavian HYPO-RT-PC trial comparing extreme hypofractionation with 7 x 6.1 Gy and conventional 39 x 2 Gy. The study has now completed inclusion of 1200 intermediate-risk prostate cancer patients.

A number of cancer centers and cooperative trial groups have decided to change to a moderately fractionation schedul as standard of care in radiation therapy for localized prostate cancer. Moderate hypofractionation seems to be associated with acceptable morbidity and seems safe with use of modern highly precise radiation technologies. It is cost-effective, convenient and competetive to robotic assisted prostatectomy, but findings from clinical trials do not consistantly support a high fractionation sensitivity of prostate cancer.

5. Reference List

[1] Brenner D J and Hall EJ 1999 *Int. J. Radiat. Oncol. Biol. Phys.* **43** 1095-101
[2] Miralbell R et al 2012 *Int. J. Radiat. Oncol. Biol. Phys.* **82** e17-24
[3] Fosså S D *et al* 2016 *Eur. Urol.* **70** 684-91
[4] Petersen S E, Bentzen L *et al* 2014 *Radiother. Oncol.* **111** 94-9
[5] Lukka H *et al* 2005 *J. Clin. Oncol.* **23** 6132-8
[6] Yeoh E E *et al* 2011 *Int. J. Radiat. Oncol. Biol. Phys.* **81** 1271-8
[7] Arcangeli G *et al* 2010 *Int. J. Radiat. Oncol. Biol. Phys.* **78** 11-8
[8] Pollack A *et al* 2013 *J. Clin. Oncol.* **31** 3860-8
[9] Dearnaley D *et al* 2015 *Eur. J. Cancer* **51** S712
[10] Lee W *et al* 2015 *Int. J. Radiat. Oncol. Biol. Phys.* **94** [Suppl]
[11] Incrocci L *et al* 2015 *Int. J. Radiat. Oncol. Biol. Phys.* **94** [Suppl]
[12] Wilkins A *et al* 2015 Lancet Oncol. **16** 1605-16
[13] Borre M *et al 1998* APMIS **106** 389-95

SBRT for recurrent head and neck cancer

M Garg, R Kabarriti, S Baliga, C Guha, W Tome, S Kalnicki
Montefiore Medical Center, Albert Einstein College of Medicine, New York, USA

Email: mgarg@montefiore.org

Abstract. The management of patients with recurrent head and neck cancers is complex. Concerns over toxicity with re-irradiation have limited its use in the clinical setting. Stereotactic Body Radiation Therapy (SBRT) has emerged as a highly conformal and precise type of radiotherapy and has the advantage of sparing normal tissue. Although SBRT is an attractive treatment modality, its use in the clinic is limited, given the technically challenging nature of the procedure. In this review, we attempt to provide a comprehensive overview of the role of re-irradiation in patients with recurrent head and neck cancers, with particular attention to the advent of SBRT and its use with systemic therapies such as cetuximab.

1. Introduction

SBRT is not only more convenient as it reduces the overall duration of treatment, but it also exploits alternate mechanisms of cell kill and tumor control over traditional fractionated radiotherapy, which might be highly relevant for recurrent disease. Results of SBRT have been highly encouraging in several malignancies including lung, spine, and liver tumors [1-6]. However, use of SBRT in head and neck cancers has been guarded because of several reasons: Head and neck has a complex anatomy with several radiosensitive structures. Tolerance to high doses per fraction of SBRT is still not known for these structures. Conventional fractionated radiation therapy as a primary modality in head neck cancers is quite successful in providing disease control and increasing the dose per fraction might result in higher late toxicities. Furthermore, several head neck cancers metastasize to neck lymph nodes, which have been traditionally treated electively. Including elective nodal radiation in an SBRT field can potentially increase the likelihood for toxicity.

2. Studies reporting SBRT for recurrent HN Cancers

The Role of SBRT in head and neck cancers is evolving with investigators studying to find the appropriate patients and dose fractionation schedule. There have been reports of doses ranging from 6 Gy to 9 Gy per fraction over 5-6 treatments (Table 1). These studies are not only aimed at providing palliation, but also long term LC. A report from Henry Ford (2009) used 30-38 Gy over 5-6 fractions and reported a RR of 77% and complete response (CR) rate of 31% [7]. Cegniz et al., demonstrated a RR of 58% with 27% of patients with stable disease (SD) that did not progress with SBRT [8].

2.1 Tumor Volume

Kodani et al. reported a 2-year OS of 50% and a relative response of 61% for 21 patients who were re-irradiated with head and neck tumors. The study also demonstrated that overall survival was better in patients with a smaller target volume. In this study, the median tumor volume irradiated was 11.6 cc

[9]. In another study, Vargo et al., analyzed 34 patients with pathologically proven recurrent, non-squamous cell cancers of the head and neck (NSCHNs) who underwent re-irradiation with SBRT to a median dose of 40 Gy in 5 fractions and found that local control was significantly improved for tumors <25 ml [10].

2.2 Toxicities

Siddiqui et al., treated 65 patients with SBRT for recurrent head and neck cancer and demonstrated that 4 patients developed grade 4 toxicities, all of whom were previously irradiated. These toxicities included 3 patients with fistulas and one with ulceration [7]. Vargo et al., reported acute and late grade 3 toxicities in 6% of patients respectively and there were no grade≥4 toxicities [10]. Kodani et al., reported that 6/21 patients who were previously irradiated developed grade IV late complications, with two dying from carotid blow out syndrome. In both of these patients, the site of disease recurrence was the cervical lymph nodes and the maximum dose to the carotid artery was 30.7 Gy and 31.7 Gy, respectively [9]. In another study by Cegniz et al., 46 patients were treated with a SBRT technique (median dose of 30 Gy) for recurrent unresectable and previously irradiated head and neck cancer. They reported a CBOS rate of 17% [8]. The authors also noted that carotid blow out occurred in patients whose tumor surrounded half or more of the carotid artery wall and when the arteries received a 100% of the prescribed re-irradiation dose.

3. Systemic therapy combined with SBRT for locally recurrent tumors

Heron et al., performed a single institution matched case control study of patients treated with SBRT alone (n=35) or SBRT with weekly cetuximab infusion during SBRT (n=35) (8). A complete response was obtained in 34.3% of patients who underwent SBRT alone compared to 45.7% in patients who underwent combined treatment with cetuximab. The 1 and 2 year LC rates were 53.8% and 33.6% for matched SBRT alone patients and 78.6% and 49.2% respectively for patients undergoing combined treatment (p=0.009). Of even more important significance, the 1- and 2 year OS rates were 52.7% and 21.1% for SBRT only and 66% and 53.5% in the combined arm (p=0.031). There were no grade 4 or 5 higher toxicities observed and no significant difference in grade 1-3 events between groups [11]. In a follow up study, 50 patients were enrolled and received concurrent cetuximab (400 mg/m2 on day-7 and then 240 mg/m2 on days 0 and +8) with SBRT to a dose of 40-44 Gy in 5 fractions on alternating days. The 1 year local PFS was 60%, loco-regional PDF was 37%, and distant PFS was 71%. The 1 year OS was 40% with a median OS of 10 months. The treatment was well tolerated with acute and late grade 3 toxicity observed in only 6% of patients [10]. Comet et al., performed a Phase I feasibility study for combined SBRT and cetuximab for locally recurrent HNC [12]. Overall, 40 patients were prospectively treated and 15 patients received concomitant cetuximab. Of the patient who received concomitant cetuximab, the overall response rate was 75% ,2 patients experienced a skin rash with cetuximab and 3 experienced grade 3 toxicities such as mucositis, dysphagia, induration and fibrosis. Lartigau and colleagues performed a Phase II trial of 56 patients with inoperable recurrent or new primary tumor in a previously irradiated area to a dose of 36 Gy in 6 fractions and concomitant cetuximab beginning with a test dose of 400 mg/m2 the week before SBRT [13]. During the two weeks of SBRT and in the following 2 weeks, patents received a weekly injection of 250 mg/m2. The one-year OS rate was 47.5%. The overall objective best response rate was 69.4%. A complete response was noted in 49% of patients, partial response in 20.4%, and stable disease in 22.5% of patients. Only 8.2% of patients had progressive disease. 18 patients had grade 3 toxicity and there was one toxic death from hemorrhage.

4. Planning/Physics Considerations

Proper patient selection is paramount for achieving good tumor control and minimizing toxicity in these patients. Factors that influence patient selection include tumor size or volume, location of the tumor, re-irradiation schedule/dose, and other co-morbid conditions such as diabetes mellitus, which may impair proper wound healing. Tumor delineation using co-registration with PET-CT or/and MRI is mandatory. Several studies have found that PET/CT imaging changes treatment volumes in a significant portion of patients. Treatment volumes were smaller in patients after PET/CT was employed in several series [14, 15]. In tumors of base of skull, sinuses and nasopharynx, MRI based contours have a high likelihood of reducing target volumes. SBRT studies have found a correlation between tumor volume and local control with 25-30 cc appearing as a cut- off point. However, Rwigema et al, 2011 showed that dose escalation to 50 Gy/5 fractions for patients with GTV>25 cc resulted in comparable SBRT response rates and toxicities to those with GTV<25 cc [16]. Location of the tumor is also another critical component of patient selection. Tumor invading into rosenmueller's fossa or/and foramen lacerum, tumor involving more than half the circumference of the carotid and diffuse mucosal involvement/ulceration are all unfavorable characteristics which could cause higher rate of late toxicity. Another important consideration to minimize late toxicity such as carotid blow out syndrome (CBOS) is the radiation schedule. Yazici et al., demonstrated that every other day SBRT protocol compared to daily SBRT resulted in improved CBOS free median OS [17]. Furthermore, CBOS did not occur in any patients with a maximum carotid artery radiation dose <34 Gy. The planning process including treatment field design is also very important. Non-opposing non-coplanar beams are preferable and typically >10 beams of radiation should be used with roughly equal weighting. Because of uncertainties in beam commissioning resulting from electron disequilibrium within small beam apertures, a minimum field size of 3.6 cm is recommended except when the entire PTV is not covered in the beam. It is also important to remember to prescribe to the 80% isodose line, as the dose fall off from the prescription isodose to half the prescription dose typically occurs over the shortest distance if the dose is prescribed to the 80% isodose shell. RTOG SBRT protocol dose conformity guidelines for other sites can be adapted as general rules for target coverage and dose fall-off in H&N SBRT as well, while respecting organs-at-risk doses

5. Conclusions

In summary, SBRT represents an attractive treatment modality for recurrent, previously irradiated unresectable head and neck tumors. The studies reported so far have shown low toxicity rates, good overall response rates, with potential for improved local control. Larger multi-institutional studies are needed to further identify the ideal patient population and to explore appropriate dose-fractionation schedule.

6. References

[1] Chang E L *et al* 2007 J. Neurosurg. *Spine* **7** 151-60
[2] Fakiris A J *et al* 2009 *Int. J. Radiat. Oncol. Biol. Phys.* **75** 677-82
[3] Gerszten P C *et al* 2007 *Spine (Phila Pa 1976)* **32** 193-9
[4] Timmerman R *et al* 2003 *Chest* **124** 1946-55
[5] Timmerman R *et al* 2010 *JAMA* **303** 1070-6
[6] Tse R V *et al* 2008 *J. Clin. Oncol.* **26** 657-64
[7] Siddiqui F *et al* 2009 *Int. J. Radiat. Oncol. Biol. Phys.* **74** 1047-53
[8] Cengiz M *et al* 2011 *Int. J. Radiat. Oncol. Biol. Phys.* **81** 104-9
[9] Kodani N *et al* 2011 *J. Radiat. Res.* **52** 24-31
[10] Vargo J A *et al* A 2015 *Int. J. Radiat. Oncol. Biol. Phys.* **91** 480-8
[11] Heron D E *et al* 2011 *Am. J. Clin. Oncol.* **34** 165-72
[12] Comet B *et al* 2012 *Int. J. Radiat. Oncol. Biol. Phys.* **84** 203-9
[13] Lartigau E F *et al* 2013 *Radiother. Oncol.* **109** 281-5
[14] Deantonio L *et al* 2008 *Radiat. Oncol.* 3 29
[15] Moule R N *et al* 2010 *Radiother. Oncol.* **97** 189-93
[16] Rwigema J C *et al* 2011 *Am. J. Clin. Oncol.* **34** 372-9
[17] Yazici G *et al* 2013 *Radiat. Oncol.* **8** 242

Table 1. Summary of Re-irradiation studies using SBRT

Study	N	Follow up (Months)	Median TV	Fractions	Total Dose (Range)	Response	Toxicity
Roh 2009	36	17.3	22.6 cc	3-5	18-40 (median 30)	RR:80% 2 yr OS 30.9%	Acute G3-50% ST necrosis: 8%
Siddiqui 2009	36	NA	15.5 cc	5-6	30-38 Gy (18-40)	CR:31% 1 yr OS: 60% MS: 6.7 M	Late Grade 4: 8%
Heron 2010	85	19.4	25 cc	1-5	35 Gy (15-44)	RR 68% 2 yr OS 16.1% 2 yr LC 30.7%	Late Grade 3: 4%
Cengiz 2011 (Turkey)	55	NA	NA	1-5	30 Gy (18-35)	RR 58% 1 yr OS 46% MS 10.5 Mo	Late G2 or greater: 13.3% CBO: 17.8%
Unger 2010	65	26	75 cc	5	30 Gy	2 yr OS 33% MST 11.5 M	Late Grade 4: 9%
Voynov 2007	22	19	19.1 cc	NA	24 (10-36)	2 yr OS: 22% 2 yr LC: 26% MST: 12 M	No grade 4 or 5 toxicity
Himei 2003 (Japan)	31	16.3	41.2 cc	1-6	30 Gy (15-40.3)	RR: 74%	Acute mucosiits: 15%
Ogita 2009	58	NA	31.8 cc	3-8	31 Gy	2 yr OS: 29.2% MS 15.5 M	Ulcer: 414%
Kodani 2011 (Japan)	34	51 M	10 cc	3-8	30 Gy(19.5-42)	RR 61% 2 yr OS: 50%	Severe late: 28%

Fast dose analysis of movement effects during treatments with scanned proton and carbon-ion beams

A Vignati[1], M Varasteh Anvar[1,2], S Giordanengo[1], V Monaco[1,2], A Attili[1], M Donetti[3], F Marchetto[1], F Mas Milian[4], M Ciocca[3], G Russo[1], R Sacchi[1,2] and R Cirio[1,2]
[1]Istituto Nazionale di Fisica Nucleare (INFN), Torino, Italy
[2]Università degli Studi di Torino, Torino, Italy
[3]Centro Nazionale di Adroterapia Oncologica (CNAO), Pavia, Italy
[4]Universidade Estadual de Santa Cruz, CNPq Fellow, Bahia, Brazil

E-mail: anna.vignati@to.infn.it

Abstract. Charged particle therapy delivered using scanned pencil beams shows the potential to produce better dose conformity than conventional radiotherapy, although the dose distributions are more sensitive to anatomical changes and patient motion. Therefore, the introduction of engines to monitor the dose as it is being delivered is highly desirable, in order to enhance the development of adaptive treatment techniques in hadrontherapy. A tool for fast dose distributions analysis is presented, which integrates on GPU a Fast Forward Planning, a Fast Image Deformation algorithm, a fast computation of Gamma-Index and Dose-Volume Histogram. The tool is being interfaced with the Dose Delivery System and the Optical Tracking System of a synchrotron-based facility to investigate the feasibility to quantify, spill by spill, the effects of organ movements on dose distributions during treatment deliveries with protons and carbon-ions. The dose calculation and comparison times for a patient treated with protons on a 61.3 cm^3 planning target volume, a CT matrix of 512x512x125 voxels, and a computation matrix of 170x170x125 voxels are within 1 s per spill. In terms of accuracy, the absolute dose differences compared with benchmarked Treatment Planning System results are negligible ($<10^{-4}$ Gy).

1. Introduction

Although showing a strong expansion in recent years [1], charged particle therapy as a standard treatment is still developing, and auxiliary technologies that are standard in modern photon radiotherapy [2, 3] systems are still lacking [4]. In particular, adaptive radiotherapy methods would need to be introduced, since the advantage of the better dose conformity offered by charged particle therapy comes to the price of greater sensitivity to delivery-time targeting uncertainties [5], such as patient motion. This work presents a tool (named RIDOS, from *Real-Time Ion DOse Planning and Delivery System*) for fast dose distribution analysis, able to monitor the dose as it is being delivered, whilst maintaining similar accuracy to current clinical standard and without prolonging the delivery time. The core of the tool is represented by a benchmarked irradiation-outcome computation algorithm (*Forward Planning*, FP) for scanned ion beams [6], parallelized and adapted to run efficiently on the GPU architecture (*Fast-FP*, F-FP). The tool receives in real-time the measured beam parameters through a direct and transparent connection with the Dose Delivery System (DDS) of a synchrotron-based facility. Online motion monitoring data will be used to correlate tumour and organ motion with the temporal structure of the beam delivery and 4D computed tomography (CT) image to reconstruct a dose-time distribution, incorporating both interplay effects and range changes. The dose distribution will be repeatedly calculated during the irradiation, i.e. after the delivery of each energy layer (spill). Moreover, thanks to a Fast Image Deformation (F-ID) algorithm, each spill-dose is mapped back to the reference image. This would make possible to monitor the progressive emergence of a motion-corrected dose distribution during treatment, and therefore to promptly identify motion artefacts.

2. Materials and Methods

The scheme of the RIDOS pipeline is shown in Fig. 1, where the reported times are related to a synchrotron-based facility. Some of the tool components are executed before the start of the treatment, while the others are performed in the inter-spill time, which is approximately 3 seconds. The planned and the delivered dose are calculated in around 300 ms, while the fast dose comparison takes 400 ms. The whole pipeline is completed by far within 1 second.

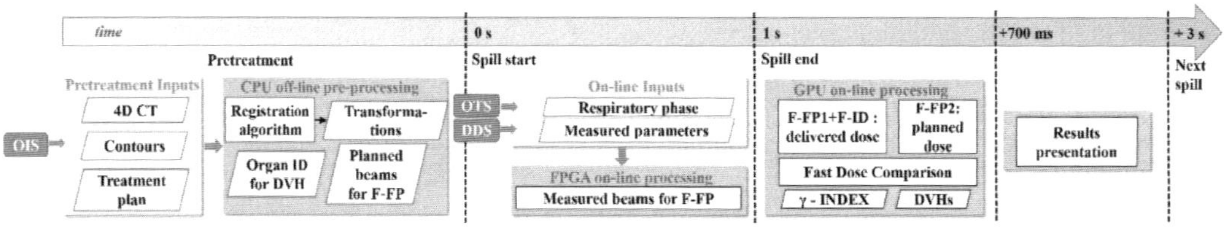

Figure 1. General scheme of the presented tool. OIS = Oncological Information System; OTS = Optical Tracking System; DDS = Dose Delivery System; F-FP = Fast Forward Planning; F-ID = Fast Image Deformation.

2.1 CPU off-line pre-processing

2.1.1 Registration. A 4D-CT scan is an image covering the entire breathing cycle, composed by a set of 3D-CT volumes, each corresponding to a particular breathing phase. Due to the respiration, voxels may migrate and distort among breathing phases, making the comparison of the different 3D-CT volumes on a voxel-by-voxel basis problematic. Image registration is a method of aligning two images into the same coordinate system, so that the aligned images can be directly compared, combined and analyzed [7]. A deformable image registration method was performed on CPU, before the treatment, between each breathing phase CT (moving image) and the CT used to plan the treatment (reference image), in order to obtain the corresponding transformation. A parametric non-rigid registration method was adopted, using Bspline transformation and Mattes Mutual Information [8] as similarity measure. Those transformations are then used by the F-ID algorithm (see paragraph 2.3.1).

2.1.2 Pretreatment inputs. The following data are loaded and processed before the treatment start: a) the patient CT (or 4D-CT), needed by the off-line registration step, by the F-FP, and the F-ID; b) the target and the region-of-interest contours, to compute the Dose Volume Histogram (DVH); c) the treatment plan with the list of planned beam parameters to compute the reference dose distributions, spill by spill.

2.2 FPGA on-line processing

A National Instruments (NI - Austin, USA) digital I/O module, equipped with a Xilinx Virtex II FPGA (San Jose, California, US) and hosted on a NI-PXI chassis, interfaces the DDS with the Workstation (WS). The FPGA firmware receives in real-time the measured beam parameters (i.e. fluence and positions) as soon as they are available in the DDS FPGA. These data are transferred spot by spot to the PXIe-CPU via DMA-FIFO to be promptly sent to the WS. A second NI-FPGA module, in the same NI-PXIe system, will receive on-line the three-dimensional patient position corrections and the respiration phase from the optical tracking system (OTS).

2.3 GPU on-line processing

2.3.1 Fast Forward Planning and Fast Image Deformation. The F-FP algorithm has been obtained by implementing on GPU the computing kernel of a benchmarked Treatment Planning System (TPS) called PlanKIT [6]. The dose calculation program, implemented in the CUDA (Nvidia Corporation, Santa Clara, CA, USA) C/C++ programming language, consists of the following steps:

- ray-tracing operations are performed in the voxelized CT volume to identify the segments of the beam axis belonging to different traversed materials;
- the path length of a beam in a heterogeneous material is converted into its water equivalent path length (WEPL), using the information of the stopping power assigned to each CT voxel by a CT-scanner-specific curve that relates Hounsfield numbers to stopping power ratios;
- a downsampled matrix of the CT matrix (*computing grid*) is hereinafter considered, and for each beam the voxels of the computing grid are selected within a chosen radial distance (*radial cut-off*) from the beam axis. The radial cut-off is the distance at which the interaction is considered.
- for each voxel position $r_m=(x_m,y_m,z_m)$ in the beam coordinate system, the projection on the beam axis z_m is rescaled to z_w according to the WEPL and a water-equivalent position $r_w=(x_m,y_m,z_w)$ is obtained;
- the dose for every selected voxel is obtained through the interpolation from the look-up tables (LUTs) of PlanKIT. The LUTs are 3D sampling of the beams effectiveness in water, in terms of dose per single particle, obtained simulating the used beamline with the Monte Carlo method (Fluka).

Spill by spill, the F-FP calculates the dose on the specific CT phase and on the reference CT. The cumulative planned dose is repeatedly updated by summing the dose calculated on the reference CT. The F-ID receives in input the transformation parameters of each CT phase created by the registration algorithm in the pretreatment step, and warps the dose distribution of the specific CT phase according to the reference CT, and the cumulative delivered dose is updated.

2.3.2 Fast dose comparison. The delivered dose is compared online to the planned dose distribution using a fast 3D gamma evaluation algorithm. The method was implemented in CUDA according to Persoon et al [9]. Moreover, the DVHs are calculated.

3. Results

A WS HP Z420 (2xIntel XeE5-2670v2) equipped with a NVIDIA Tesla K20c was used to test the algorithms. Since, at present, the connection with the DDS and the OTS is not complete, to evaluate the results of the F-FP, F-ID and Fast Dose Comparison an artificial rotation of 2 degrees (around the perpendicular to the axial plane) was applied to a static CT case, and the rotated CT image was used as CT phase. The static CT case was referred to a 61.3 cm^3 planning target volume (PTV) brain case, treated with protons (one beam entrance direction, 39 energy layers, and 1248 spots). The dimension of the CT matrix was 512x512x125 voxels (320x320x250 mm^3), while the computing grid was 170x170x125 voxels. Table 1 presents the time gain of the RIDOS F-FP in respect to the PlanKIT FP to calculate the dose distribution of the entire treatment, although it is worth to underline that no particular effort has been made to optimise the computing times of PlanKIT.

The gamma index calculation times were inversely dependent on the dose interpolation resolution. Using an interpolation of 0.1 mm and the already defined computing grid, gamma values were obtained in about 400 ms. Fig. 2 presents the results of the dose comparison step. The absolute differences between the dose distribution computed by the RIDOS F-FP and the PlanKIT FP are negligible ($<10^{-4}$ Gy). As already mentioned, for individual spills, the F-FP dose calculation and the F-ID times are about 300 ms, while fast gamma index and the DVHs computation times are estimated to be around 400 ms. Therefore, the dose computation and comparison of one spill are expected to be within 1 s. Similar results are obtained considering a carbon-ion treatment.

Table 1. Computing times of the RIDOS F-FP and the PlanKIT FP, and related gain, by varying the radial distance from each beam axis at which the interaction is considered (*radial cut-off*). The dose is calculated for the number of voxels selected, which increase by augmenting the radial cut-off.

Radial cut-off (mm)	Number of voxels	RIDOS F-FP time (s)	PlanKIT FP time (s)	Gain
20	44678	1.35	157	116
40	178702	2.33	636	273
50	279243	2.76	997	361
80	714861	4.68	2534	541

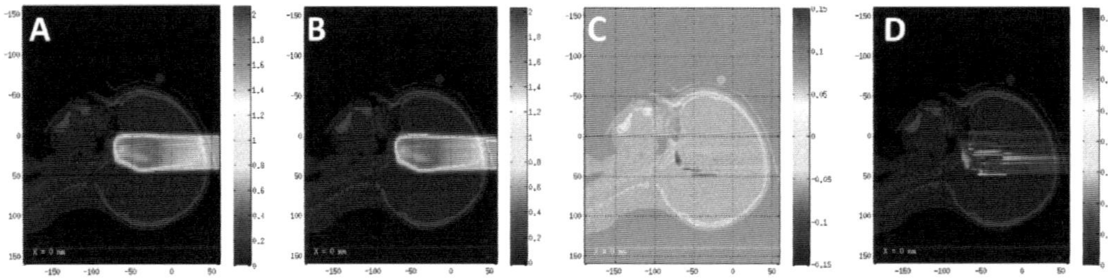

Figure 2. Comparison of the planned dose calculated on the reference CT in Gy (A), the delivered dose calculated on the rotated CT in Gy (and that warped according to the reference CT by the F-ID algorithm) (B), the absolute dose difference between A and B in Gy (C), and the gamma index (C) for a distance to agreement = 3 mm, and dose difference = 3%.

4. Discussion

The precise dose localization enabled by ions makes them favourable for highly conformal radiotherapy treatments but also sensitive to uncertainties due to range variations, patient intra-fractional motion and interplay effects between target and beam movement during delivery. This work represents an important step towards adaptive hadron therapy providing the pre-clinical validation of a new GPU-based system, able to monitor the dose as it is being delivered. These preliminary results suggest the readiness of the tool for the possible test in a clinical environment, to verify its potential contribution to the developments of on-line dose measurements and verification modalities. Additionally, the system could find application in the real-time interactive treatment planning [10], or to monitor the dose delivery in hypofractionated treatments.

5. Acknowledgments

The RIDOS project was founded by the INFN-CSN5 Grant, Young Researcher 2013.

6. References

[1] PTCOG (Particle Therapy Co-Operative Group) 2014 (http://ptcog.web.psi.ch/)
[2] Vial P *et al* 2008 *Med. Phys.* **35** 1267-77
[3] Hill R *et al* 2010 *Med. Phys.* **37** 4355-63
[4] da Silva J *et al* 2015 *Phys. Med. Biol.* **60** 4777
[5] Ammazzalorso F *et al* 2014 *Radiat. Oncol.* **9** 279
[6] Russo G *et al* 2016 *Phys. Med. Biol.* **61** 183
[7] Kadoya N 2014 *J Radiol. Radiat. Ther.* **2** 1042
[8] Mattes D *et al* 2003 *IEEE T-MI* **22** 120
[9] Persoon L C G G *et al* 2011 *Med. Phys.* **38** 4032
[10] Otto K 2014 *Phys. Med. Biol.* **59** 4845

On the dose to a moving target in stereotactic ablative body radiotherapy to lung tumors

V Feygelman, T J Dilling, E G Moros, G G Zhang
H. Lee Moffitt Cancer Center
Tampa, Florida, 33612, USA

E-mail: vladimir.feygelman@moffitt.org

Abstract. This review summarizes the hierarchy of potential dose inaccuracies in lung SABR in terms of their expected clinical impact. The two main terms are targeting accuracy and adequacy of the dose calculation algorithm. One can associate dose-errors at the 50-100% (zero order) and 10-20% (first order) levels with the former and the latter, respectively. At the first order level, strong evidence exists that using dose algorithms which do not account for 3D density scaling is associated with diminished local control. On the other hand, the second-order target dose-errors due to either static approximations to full 4D calculations, or interplay during modulated delivery, are rather unlikely to rise above 5% (conservatively, $\leq$ 1% tumor control probability change).

1. Introduction

To be effective, precision radiotherapy must take target motion into account. Motion management is even more important in stereotactic ablative body radiotherapy (SABR), where hypofractionated high dose is delivered with tight margins to small targets [1]. The SABR in lung poses particular dosimetric challenges because of the often substantial differences in density between the tumor and surrounding normal lung tissue. This leads to the loss of commutative invariance between the fluence/motion and fluence/dose convolution steps [2], thus requiring, at least in theory, full four-dimensional (4D) dose calculations with deformable image registration (DIR) [3]. In this review, we attempt to systematically quantify the effects of various simplifications and approximations to this ideal 4D dose calculation, and correlate them with clinical outcomes.

2. Dose-errors hierarchy

2.1. Geographical miss

The first and obvious effect of the patient or target motion is the possibility of a blatant geographical miss. We will consider it self-evident that poor targeting (>50% dose-error to at least a portion of the tumor) would lead to inferior clinical outcomes. Listed in the order of sophistication, which also roughly corresponds to the chronological order of introduction, the motion management methods include: external immobilization and patient position monitoring; tumor motion assessment at simulation (fluoroscopy or 4D CT) and its incorporation into the internal and planning target volumes (ITV and PTV); daily image guidance (IGRT); surrogate-based gating; and real-time direct image-guidance / tumour tracking [4, 5]. We will consider adequate targeting the zero-order term of the dosimetric accuracy hierarchy, similarly to an expansion series.

2.2. Calculation algorithms: correlation with SABR outcomes

The next (first-order) term in our accuracy construct is the fidelity of the calculation algorithm. The challenges of dose calculations in low-density tissue such as lung are well documented [6]. It is furthermore established that the only algorithms producing reasonably accurate results in lung are those explicitly taking the lateral change in tissue density (and resulting changes in the dose-deposition kernel) into account ("type b" models by Knöös et al [7]). While it was clear that without the inhomogeneity corrections, or with only 1D correction, the planned dose was misleading, in the pre-SABR era direct clinical correlation between the accuracy of the TPS algorithm and local control was not particularly easy to establish. It was due partly to the generally poor tumor control, and partly to the fact that the cooperative groups were using dose metrics and monitor unit settings based on the homogeneous datasets, extending this practice even into the earlier SABR studies [8]. In retrospective analysis by Xiao et al [8], after recalculating the unit-density plans with the heterogeneity-corrected convolution/superposition algorithms for the RTOG 0236 SABR protocol, the PTV volume receiving the prescription dose decreased on average by 10%.

The advent of SABR allowed for better direct correlation of the TPS accuracy with local control. A confluence of several factors is responsible for that. Local control rates are generally high (i.e. weighted average of 90% for Stage I from the review by Solda et al [9]), and failures were therefore visible and piqued the researchers' interest. As growing evidence of the dose-response relationship was being reported, it was not lost on the interested physicists that a small near-unit density tumor in the middle of the low density lung tissue was an ideal setup for the 1D density-correction algorithms to fail rather dramatically ([10] is just one of many examples). In addition, it was reasonably easy to geographically correlate local failures with dosimetric deficiencies [11, 12].

A study by Liu et al [12] of 77 patients with median follow-up of 13.4 mo found that with all other parameters being equal, the patients treated with the CyberKnife based on the equivalent path length calculations had diminished local control compared to the Monte Carlo-calculated cohort treated to the same nominal biologically equivalent doses (BEDs). A larger study by Latifi et al [10] involved 201 patients (23 mo median follow-up) whose plans were calculated with either Pencil Beam (PB) or Collapsed Cone Convolution (CCC) algorithms to the same nominal physical dose (10 Gy×5). Crude local recurrence rates were 21.5% vs. 4.7% in the PB and CCC cohorts, respectively, and the hazard ratio difference was statistically significant by Gray's competing risks analysis. The PB cohort was treated very similarly to the prior smaller study by Videtic et al [13], yet demonstrated diminished local control (78 vs. >93%). While not definitively explained, the comparison results are consistent with the idea that a dose regimen of 50 Gy in 5 fractions to 95% of the PTV is right at the cusp of the BED adequate for local control (100 Gy, [14, 15]).

Based on the combination of dosimetric correlations in the cited papers, we will conservatively assume that the suboptimal planning algorithm introduces a clinically significant dosimetric error of $\geq$10%, and that will quantify the first-order term in the error hierarchy.

2.3. Second-order effects

The remaining task is to establish the clinical significance of the second-order terms in our dose-error expansion series. The most straightforward approach is to find the gradient of the tumor control probability (TCP) curve, and compare it to the magnitude of dose-errors characteristic of the second-order effects. To that end, we numerically differentiated three original and compiled [12, 16] TCP vs. BED ($\alpha/\beta = 10$) curves. The steepest derivative at 100 Gy BED showed a ~0.2% change in TCP per 1 Gy (1%). Therefore a

5% target underdosage corresponds to 1% TCP loss, which we will conservatively consider the threshold of clinical significance around 100 Gy BED. The threshold will be even higher for the more robust dose-fractionation schemes [17]. The question then becomes if the effects of various approximations to the full 4D calculations, and/or possible interplay consequences, could rise above the threshold of clinical importance.

2.3.1. Approximations to full 4D calculations. The first step is elimination of the dose calculation on every phase, which could be time-consuming, particularly for VMAT treatments. Glide-Hurst et al [18] established feasibility of this approach by performing the dose calculation on an average CT, while still using DIR for target dose accumulation throughout the breathing cycle. The largest singular discrepancy between the simplified and full 4D calculations in the relevant gross tumor volume (GTV) dosimetric indices was -2.4% (D_1 in one plan), with the vast majority well under 1%. While DIR software packages are relatively widely available by now, the method introduces additional steps and its own uncertainties [19]. The next step, elimination of DIR, was explored by Rao et al [20]. With a simple 3D calculation on a single respiratory phase, the average and maximum differences from a full 4D approach for the GTV D_{99} did not exceed 1.2 and 2.4%, respectively. Similar conclusions were reached by Admiraal et al [21]. Both the 3D and 4D dose calculations on a deformable phantom agreed with measurements in the GTV, and the differences were observed primarily at the edges of the PTV [22]. That error can be further remedied by intelligent CT density overrides [23]. A somewhat different approach to circumventing the need for DIR in lung was examined by Zhang et al [24]. It relies on the concept of motion-weighted clinical target volume (mwCTV), which carries more 4D information into planning than the 3D internal target volume (ITV) method. This solution allows, in principle, plan optimization based on occupancy weighting of the CTV, and generation of motion-weighted DVHs (mwDVH) that approximate the DVHs (but not the dose distribution) of the full 4D dose accumulation. The percentage of the CTV volume covered by the prescription dose differed on average by only 0.3±0.7% between the mwDVH and full 4D calculations, which was statistically less than for 3D/ITV calculations.

2.3.2. Interplay. In brief, point dose measurements [25] tend to over-emphasize the importance of the interplay between target motion and radiation fluence modulated in space and time, while full 3D calculations [20] or measurement-guided reconstruction [26] indicate that the important target DVH parameters are not substantially affected. Overall, the magnitudes of target dose-errors due to either 3D calculation approximations or interplay during delivery are rather unlikely to rise above 5% (1% TCP change).

3. Conclusions
Clinical evidence exists that using less accurate, 1D heterogeneity corrections is associated with diminished local control in lung SABR. The second order effects, such as static approximations to 4D dose calculations and potential interplay during modulated deliveries, do not appear likely to affect local control.

4. References
[1] Benedict S H *et al* 2010 *Med. Phys.* **37** 4078-101
[2] Chetty I J *et al* 2003 *Med. Phys.* **30** 1776-80
[3] Guckenberger M *et al* 2007 *Int. J. Radiat. Oncol. Biol. Phys.* **69** 276-85
[4] Hill R *et al* 2010 *Med. Phys.* **37** 4355-63
[5] Vial P *et al* 2008 *Med. Phys.* **35** 1267-77
[6] Papanikolaou N *et al* 2004 *AAPM report No.85* (Madison: Medical Physics Publishing)
[7] Knoos T *et al* 2006 *Phys. Med. Biol.* **51** 5785-807

[8] Xiao Y *et al* 2009 *Int. J. Radiat. Oncol. Biol. Phys.* **73** 1235-42
[9] Solda F *et al* 2013 *Radiother. Oncol.* **109** 1-7
[10] Chetty I J *et al* 2013 *Radiother. Oncol.* **109** 498-504
[11] Latifi K *et al* 2014 *Int. J. Radiat. Oncol. Biol. Phys.* **88** 1108-13
[12] Liu M B *et al* 2013 *Practical Radiation Oncology 3* 294-300
[13] Videtic G M *et al* 2010 *Int. J. Radiat. Oncol. Biol. Phys.* 77 344-9
[14] Guckenberger M *et al* 2009 *Int. J. Radiat. Oncol. Biol. Phys.* **74** 47-54
[15] Olsen J R *et al* 2011 *Int. J. Radiat. Oncol. Biol. Phys.* **81** e299-303
[16] Wulf J *et al* 2005 *Radiother. Oncol.* **77** 83-7
[17] Baumann P *et al* 2009 *J. Clin. Oncol.* **27** 3290-6
[18] Glide-Hurst C K *et al* 2008 *Med. Phys.* **35** 5269
[19] Samavati N *et al* 2016 *Med. Phys.* **43** 233-40
[20] Rao M *et al* 2012 *Int. J. Radiat. Oncol. Biol. Phys.* **83** e251-6
[21] Admiraal M A *et al* 2008 *Radiother. Oncol.* **86** 55-60
[22] Vinogradskiy Y Y *et al* 2009 *Med. Phys.* **36** 5000-6
[23] Wiant D *et al* 2014 *Med. Phys.* **41** 081707
[24] Zhang G *et al* 2011 *Radiother. Oncol.* **99** 67-72
[25] Court L E *et al* 2010 *Med. Phys.* **37** 5850-7
[26] Stambaugh C *et al* 2013 *Med. Phys.* **40** 091710

IOP Conf. Series: Journal of Physics: Conf. Series **777** (2017) 012028 doi:10.1088/1742-6596/777/1/012028

Spine SBRT using VMAT

L Goddard[1], W Bodner[1], N P Brodin[2], M Garg[1], A Lee[1], K Mani[1], W A Tomé[1, 2]
[1]Department of Radiation Oncology, Montefiore Medical Center, Bronx, NY 10461, USA [2]Institute for Onco-Physics, Department of Radiation Oncology, Albert Einstein College of Medicine, Bronx, NY 10461, USA

Email: wtome@montefiore.org

Abstract. Metastases of the spinal column are common amongst cancer patients with approximately 18,000 new cases in North America each year that require urgent treatment. Historically radiation therapy doses have been limited due to the proximity of the spinal cord. However as image guidance and localization techniques have improved it has become possible to deliver higher radiation doses to the tumour whilst sparing the spinal cord. This paper presents some of the techniques undertaken at our center.

1. Introduction

Metastases of the spinal column are common amongst cancer patients with approximately 18,000 new cases in North America each year that require urgent treatment [1]. Metastases are the most common group of spinal column tumours, with post mortem examination of terminal cancer patients revealing spinal metastasis in more than 70% of patients [2]. Approximately 10% of cancer patients develop symptomatic lesions, with the primary presenting symptom being pain. As such the primary goal of treatment is pain relief.

A single fraction of 8Gy has been shown to have similar efficacy to a conventional fractionated regimen of 30Gy in 10 fractions, whilst maintaining a low rate of adverse affects [3]. A single fraction treatment presents an advantage to the patient in ease of treatment, and reduced machine time required at the treatment facility. Approximately 61% of patients treated in this study experienced pain relief at 1 month post treatment. The rate of pain control may be improved by treating to higher radiation doses, however, the close proximity to the spinal cord has been the limiting factor in previous studies.

Stereotactic Body Radiation Therapy (SBRT) techniques allow the delivery of high doses of radiation precisely and accurately to the target [4, 5]. The clinical feasibility of SBRT for treating spinal column metastases was first demonstrated in 2003 [6], with targeting accuracy reported as less than 1.5mm. Dose escalation studies have shown rapid improved pain relief when treated up to 16Gy in a single fraction [7]. However only a small number of patients were treated on these studies. RTOG 0631 is a multi-institutional study, began in 2009, designed to show the clinical feasibility of SBRT for the treatment of spine metastases. RTOG 0631 also aims to investigate the efficacy of 16Gy vs. 8Gy single fraction radiation therapy for the relief of pain.

This paper presents our clinical experience using VMAT for SBRT delivery and steps taken to improve it.

2. Methods

2.1 Optimization Objectives

Montefiore Medical Center (MMC) began treating spine metastases using SBRT in January 2008, planning goals at this time were simply to achieve the best possible coverage whilst meeting cord dose constraints. Since Sept 2011 additional constraints were added according to the RTOG 0631 [8] (Spine SBRT), 0813 [9] & 0915 [10] (Lung SBRT) protocols.

0631 established detailed requirements for treatment planning, which included prescription isodose surface coverage, target dose heterogeneity, high dose spillage, spinal cord dose, low dose spillage, and critical organ dose-volume limits. Regarding the low dose spillage, 0631 only indicated, "The falloff gradient beyond the target volume extending into normal tissue structures must be rapid in all directions" without quantitative definition of the dose falloff gradient. However, both RTOG 0813 & 0915 clearly specified intermediate dose spillage criteria based on the size of planning target volumes (PTVs). Dosimetric comparison of 31 lesions treated before and 35 lesions treated after these constraints were added shows improvement in almost all aspects of treatment planning [11].

2.2 Beam Arrangement

In early 2013 a standard beam arrangement utilizing Volumetric Arc Therapy (VMAT) was introduced to simplify treatment delivery and plan QA workflows. Initially treatments used a single arc with collimator set to 45°. Two coplanar 360° arcs are now used with collimators set to 0° and 90° respectively. This minimizes collimator opening size and reduces low dose spillage as shown in figure 1. Avoidance sectors are used to skip patient shoulders and/or arms as necessary.

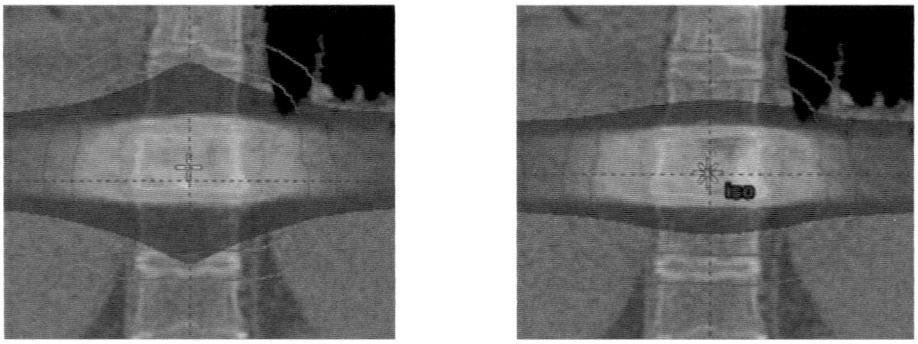

Figure 1. >5% dose colourwash for a single (left) and dual arc (right) configuration.

To show the advantages of this technique, 19 patients initially treated using Static IMRT fields were first replanned using current planning guidelines. These patients were then replanned using our standard VMAT beam arrangement. Dosimetric comparison of these plans showed a reduced maximum dose 2cm from the target (D2cm), ratio of the volume receiving 50% of prescription dose to target volume (R50) and maximum skin dose for VMAT plans, whilst maintaining target coverage and maximum dose to the spinal cord.

Additional treatment plans were also created using a 3 field Intensity Modulated Proton Therapy (IMPT) technique. As shown in Figure 2, similar cord maximum doses are achievable using IMPT, however dose conformity index (CI) for smaller targets is worse. Figure 3 shows IMPT has higher D2cm values, however, this is due to the limited number of fields used. R50 values are better for IMPT plans for larger targets.

2.3 Patient Positioning

Patient planning CT scans are performed in a head first supine position with the patient arms raised for disease of the lower thoracic spine and lumbar spine, and arms down for disease of the upper thoracic spine and cervical spine. Patients are immobilized using the Elekta BlueBAG BodyFIX vacuum cushion, to allow for comfortable stable positioning. CBCT positioning verification is performed before each treatment.

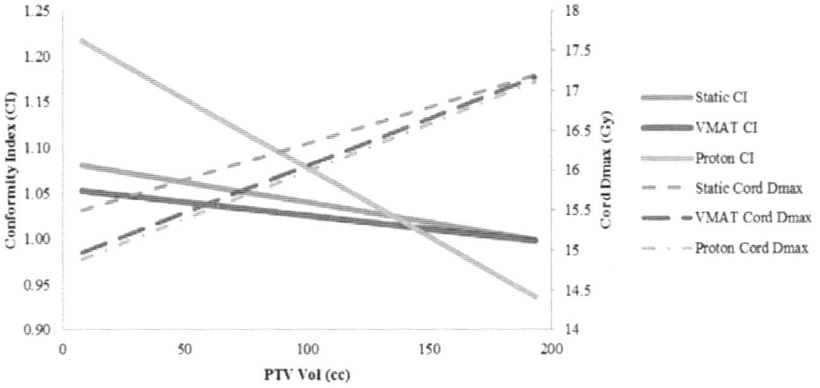

Figure 2. Conformity Index and maximum cord dose for static IMRT photon therapy, VMAT and IMPT.

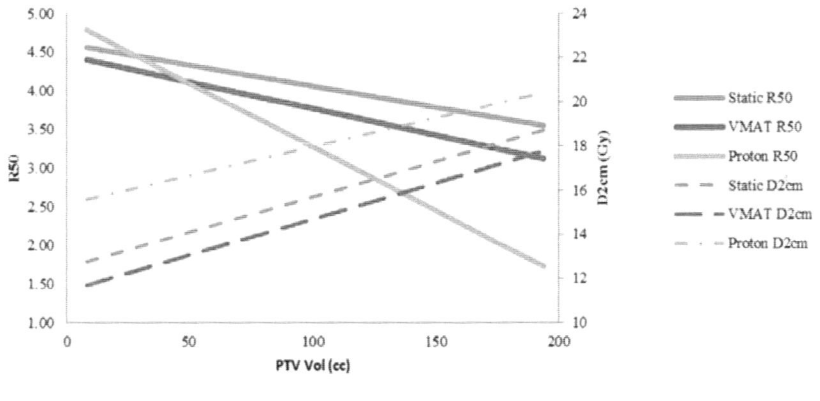

Figure 3. R50 and D2cm for static IMRT photon therapy, VMAT and IMPT.

2.4 Cord Compression

Metastatic epidural spinal cord compression (MESCC) occurs in approximately 5% of diagnosed patients [12]. If not addressed this can lead to permanent neurologic defects, including paraplegia. Conventional treatment techniques include surgery followed by radiation therapy. However for patients who are not surgical candidates definitive radiation therapy is required.

SBRT dose constraints have previously used the thecal sac as a surrogate for the spinal cord, however, RTOG 0631 recommends the true cord be delineated using an MRI registered to the planning CT. Another option for true cord delineation is CT Myelogram (CTM). Comparison of the true cord as contoured using CTM vs. MRI shows excellent agreement between the two modalities as shown in Figure 4. The use of CTM removes any uncertainties caused by image registration, and for this reason is the preferred method used at MMC.

When the true cord is within 3mm of the PTV, underdose of the PTV in this margin is acceptable to ensure Cord dose constraints are met. This is common in cases of MESCC, however, leads to sub optimal PTV coverage and the possibility of disease progression [13]. An adaptive planning study is currently underway at MMC in which patients with MESCC receive an initial SBRT treatment of 8 Gy, with the hopes of increasing the patency of the spinal canal. Patients then receive a second SBRT treatment 14-21 days later, based upon the radiographic response of the disease.

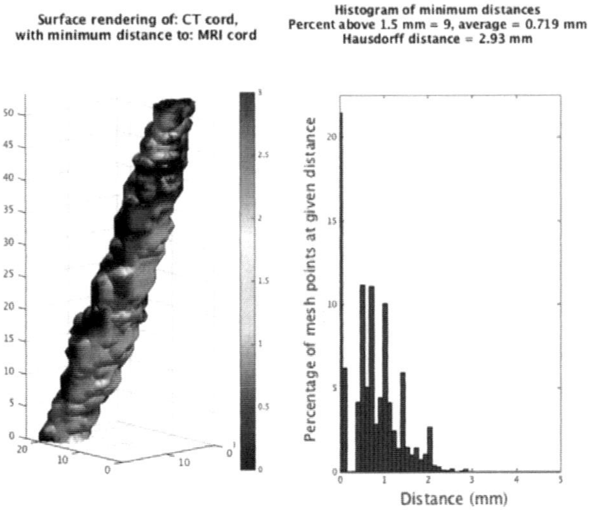

Figure 4. Distance between true cord as contoured using CTM and MRI imaging.

3. Conclusions

Using intermediate dose spillage criteria based on PTV size VMAT Spine SBRT plans can be improved regarding dose coverage, conformity, and dose falloff. In order to achieve this clearly established planning objectives are necessary to ensure quality of spine SBRT treatment plans [11].

VMAT treatment techniques offer improved dose falloff and sparing of normal tissues, whilst reducing overall treatment time compared to static field IMRT. Moreover, IMPT may offer an improved treatment in some cases and could be an important tool in the delivery of SBRT for metastatic epidural spinal cord compression. However, IMPT does not necessarily represent the optimal treatment modality unless range uncertainties can be minimized to guarantee that the dose gradient between PTV and Spinal Cord can indeed be achieved and the treatment can be safely and accurately delivered.

4. References

[1] Walsh G L *et al* 1997 *Ann. Thorac. Surg.* **64** 1611-18
[2] Perrin R G and AW Laxton 2004 *Neurosurg. Clin. N. Am.* **15** 365-73
[3] Hartsell W F *et al* 2005 *J. Natl. Cancer Inst.* **97** 798-804
[4] Hill R et al 2010 *Med. Phys.* **37** 4355-63
[5] Vial P et al 2008 *Med. Phys.* **35** 1267-77
[6] Ryu S *et al* 2003 *Cancer* **97** 2013-8
[7] Ryu S *et al* 2008 *J. Pain Symptom Manage.* **35** 292-8
[8] Ryu S *et al* 2011 *Int. J. Radiat. Oncol. Biol. Phys.* **81** S131-2
[9] R T O G (Radiation Oncology Therapy Group) 2015 *RTOG 0813 Protocol* (https://www.rtog.org/ClinicalTrials/ProtocolTable/StudyDetails.aspx?study=0813)
[10] R T O G (Radiation Oncology Therapy Group) 2014 *RTOG Protocol 0915* (https://www.rtog.org/ClinicalTrials/ProtocolTable/StudyDetails.aspx?study=0915)
[11] Hong L X *et al Med. Dosim.* **40** 181-185
[12] Cole J S and R A Patchell 2008 *Lancet Neurol.* **7** 459-66
[13] Ryu S *et al* 2010 *Cancer* **116** 2250-7

Review of gel dosimetry: a personal reflection

C Baldock
Faculty of Science, Engineering and Technology, University of Tasmania, Sandy Bay, Tasmania 7005, Australia

Email: clive.Baldock@utas.edu.au

Abstract. Gel dosimeters are manufactured from radiation sensitive chemicals which, upon irradiation, polymerize as a function of the absorbed radiation dose. These gel dosimeters have the capacity to record radiation dose distribution in three-dimensions (3D) compared to one and two-dimensional dosimeters. 3D dosimeters are radiologically soft-tissue equivalent and may be evaluated using magnetic resonance imaging (MRI), optical-computerized tomography (optical-CT), x-ray CT, ultrasound or vibrational spectroscopy.

1. Introduction

Radiation dosimetry is an important part of the radiation therapy quality assurance process in which measurements are undertaken in one, two or three dimensions to ensure that the absorbed dose, as prescribed by the radiation oncologist, is delivered correctly [1-8].

2. Fricke Gel Dosimeters

As long ago as 1950, the radiation-induced colour change in dyes was used to investigate radiation doses in gels [9]. Gel dosimetry today however, is founded mainly on the work of Gore et al who in 1984 [10] demonstrated that changes due to ionising radiation in Fricke dosimetry solutions [11] which were introduced in the 1920's, could be measured using nuclear magnetic resonance (NMR).

The nuclear magnetic resonance (NMR) relaxation properties of irradiated Fricke or ferrous sulphate dosimetry solutions show that radiation-induced changes, in which ferrous ($Fe2+$) ions are converted to ferric ($Fe3+$) ions, can be quantified using NMR relaxation measurements [11]. Further, Fricke dosimetry solutions dispersed throughout a gel matrix could be used to obtain three-dimensional (3D) spatial dose information using magnetic resonance imaging (MRI) [12]. Fricke-type gel dosimeters however do not retain a spatially stable dose distribution due to ion diffusion within the irradiated dosimeters [13, 14]. Fricke solutions with various gelling agents such as gelatine, agarose, sephadex and polyvinyl alcohol (PVA) along with chelating agents such as xylenol orange (XO) do reduce diffusion [15-17]. The diffusion problem however is considered to be a significant one in gel dosimetry.

3. Polymer Gel Dosimeters

Polymer systems for the use of radiation dosimetry were first proposed as early as 1954, where Alexander [18] discussed the effects of ionising radiation on polymethylmethacrylate. Hoecker [19] in 1958 investigated the dosimetry of radiation-induced polymerisation in liquids, and in 1961 Boni [20] used polyacrylamide as a gamma dosimeter.

A new gel dosimetry formulation was subsequently proposed based on the polymerisation of acrylamide and N,N'-methylene-bis-acrylamide (bis) monomers infused in an aqueous agarose matrix

[21]. This type of gel dosimeter, known as PAG, did not have the associated diffusion problem of Fricke gels and was shown to have a relatively stable post-irradiation dose distribution with the polymerisation reaction occurring by cross-linking of the monomers induced by the free radical products of water radiolysis [22-24]. Subsequent work investigating different compositions and formulations of polymer gel dosimeters have been summarised [25].

Although polymer-type dosimeters do not have the diffusion limitations of Fricke-type gel dosimeters, there was another significant limitation in their use. Due to the nature of their free radical chemistry, polymer gel dosimeters were susceptible to atmospheric oxygen inhibition of the polymerisation processes and, as a result, had to be manufactured in an oxygen-free environment [26].

A number of studies have been undertaken to investigate these radiological tissue-equivalent [27-29] PAG-type polymer gel dosimeters. De Deene et al [30] undertook an investigation into the overall accuracy of an anthropomorphic polymer gel dosimetry phantom for the verification of conformal radiotherapy treatments. It was established that significant issues relating to the accuracy of this dosimetry technique were a result of oxygen inhibition in the polymer gel and MRI imaging artefacts [31]. Authors continued to investigate clinical aspects of polymer gel dosimetry using MRI [32] including conformal therapy, IMRT and IMAT [33], stereotactic radiosurgery [34], brachytherapy [35], low energy X-rays [36], high-LET and proton therapy [37], boron capture neutron therapy [38].

4. Normoxic Polymer Gel Dosimeters

A significant development in the field of gel dosimetry occurred using an alternative polymer gel dosimeter formulation [39]. This new type of polymer gel dosimeter, known as MAGIC gel, bound atmospheric oxygen in a metallo-organic complex thus removing the problem of oxygen inhibition and enabling polymer gels to be manufactured on the bench-top in the laboratory. This created what was to be known as a normoxic gel dosimeter, compared with the previous PAG formulation which subsequently became known as a hypoxic gel dosimeter. The MAGIC polymer gel formulation consisted of methacrylic acid, ascorbic acid, gelatine and copper. The principal behind the MAGIC gel is in the ascorbic acid oxygen scavenger. Ascorbic acid binds free oxygen contained within the aqueous gelatine matrix into metallo-organic complexes and this process is initiated by copper sulphate. It was subsequently shown that other antioxidants could be used in the manufacture of normoxic gels [40, 41] including tetrakis (hydroxymethyl) phosphonium chloride [42]. Work has also included the development new formulations of normoxic gels [43-45] and of less toxic polymer gels [46].

With the introduction of normoxic gel dosimeters, MRI studies were undertaken to investigate their usefulness for IMRT [47], and radionuclide therapy [48].

5. Evaluation of Gel Dosimeters

Since the work of Gore et al in 1984 [10], the majority of evaluations of gel dosimeters have been undertaken with MRI. However, in 1996 the potential of optical-CT as an alternative imaging technique to MRI [49] for PAG-type polymer gel dosimeters was demonstrated [50] and further investigated [51, 52]. Subsequently, the use of X-ray CT [53-55] to image PAG-type gels was demonstrated [56]. In 1998 the use of variational spectroscopy was demonstrated to evaluate PAG-type polymer gel dosimeters [57, 58]. In 2002 the use of ultrasound to image polymer gel dosimeters was demonstrated [59-61].

6. PRESAGE™ Dosimeters

A new class of polymer dosimeter, PRESAGE™ (Heuris Pharma, Skillman, NJ) [62] was proposed in 2003 and based on clear polyurethane combined with leuco-dye leucomalachite green, an alkyl diisocyanate prepolymer and a hydroxyl reactive polyol along with a catalyst, which polymerises into optically clear polyurethane. Although not suitable for MRI evaluation, this radiation tissue-equivalent dosimeter [29, 63, 64] contains leuco dyes which have a maximum absorbance at a wavelength of 633 nm and are therefore suitable for evaluation with a He-Ne laser-based optical scanning system [65-67].

7. Other Non-radiotherapy Developments

A non-radiotherapy development in gel dosimetry was reported by Gore et al in 1997 using polymer gel dosimeters to develop an image quality test tool for MRI [68]. Hill reported the use of polymer gel dosimeters in measuring the CTDI (Computer Tomography Dose Index) on diagnostic x-ray CT scanners [69, 70] indicating the potential of using polymer gel dosimeters for diagnostic dose levels. The use of polymer gel dosimeters to measure absorbed dose from Tc-99m, the radionuclide used in nuclear medicine imaging [71].

8. Discussion and Conclusions

This paper is based on reviews on gel dosimetry previously published by the author [72-74].

9. References

[1] Hill R F *et al* 2005 *Phys. Med. Biol.* **50** N331-44
[2] Hill R *et al* 2008 *Radiat. Meas.* **43** 1258-64
[3] Vial P *et al* 2008 *Med. Phys.* **35** 1267-77
[4] Greer P B *et al* 2007 *Med. Phys.* **34** 4389-98
[5] Hill R *et al* 2010 *Med. Phys.* **37** 4355-63
[6] Hill R *et al* 2009 *Med. Phys.* **36** 3971-81
[7] Vial P *et al* 2008 *Med. Phys.* **35** 4362-74
[8] Hill R *et al* 2014 *Phys. Med. Biol.* **59** R183-231
[9] Day M J and Stein G 1950 *Nature* 166 146-7
[10] Gore J C *et al* 1984 *Phys. Med. Biol.* **29** 1189-97
[11] Fricke H and Morse S 1927 *Am. J. Roentgenol. Radium Therapy Nucl. Med.* **18** 430-2
[12] Appleby A *et al* 1986 *Med. Phys.* **14** 382-4
[13] Baldock C *et al* 2001 *Australas. Phys. Eng. Sci. Med.* **24** 19-30
[14] Harris P J *et al* 1996 *Phys. Med. Biol.* **41** 1745-53
[15] Hill B *et al* 2002 *Phys. Med. Biol.* **47** 4233-46
[16] Davies J and Baldock C 2008 *Radiat. Phys. Chem.* **77** 690-96
[17] Healy B J *et al* 2003 *Med. Phys.* **30** 2282-91
[18] Alexander P *et al* 1954 *Proceedings of the Royal Society* **A223** 392
[19] Hoecker F E and Watkins I W 1958 *Int. J. Appl. Rad. Isotop.* **3** 31-5
[20] Boni A L 1961 *Radiation Research* **14** 374-80
[21] Maryanski M J *et al* 1994 *Phys. Med. Biol.* **39** 1437-55
[22] Lepage M *et al* 2001 *J. App. Poly. Sci.* **79** 1572-81
[23] Lepage M *et al* 2001 *Phys. Med. Biol.* **46** 1061–1074
[24] Lepage M *et al* 2001 *Phys. Med. Biol.* **46** 2827-39
[25] Lepage M *et al* 2001 *Phys. Med. Biol.* **46** 2665-80
[26] Baldock C *et al* 1998 *Phys. Med. Biol.* **43** 695-702
[27] Keall P and Baldock C 1999 *Australas. Phys. Eng. Sci. Med.* **22** 85-91
[28] Venning A J *et al* 2005 *Med. Phys.* **32** 1047-53
[29] Brown S *et al* 2008 *Appl. Radiat. Isot.* **66** 1970-4
[30] De Deene Y *et al* 2000 *Magn. Reson. Med.* **43** 116-25
[31] Gustavsson H *et al* 2004 *Phys. Med. Biol.* **49** 227-41
[32] Murry P J and Baldock C 2000 *Australas. Phys. Eng. Sci. Med.* **23** 44-51
[33] Vergote K *et al* 2004 *Phys. Med. Biol.* **49** 287-305
[34] Watanabe Y *et al* 2002 *Med. Phys.* **29** 797-802
[35] Hurley C et al 2006 *Nuclear Inst. and Methods in Physics Research, A* **565** 801-11
[36] Boudou C *et al* 2004 *Phys. Med. Biol.* **49** 5135-44
[37] Gustavsson H *et al* 2004 *Phys. Med. Biol.* **49** 3847-55
[38] Gambarini G *et al* 2004 *Appl. Radiat. Isot.* **61** 759-63
[39] Fong P M *et al* 2001 *Phys. Med. Biol.* **46** 3105-13

[40] De Deene Y *et al* 2002 *Phys. Med. Biol.* **47** 3441-63
[41] De Deene Y *et al* 2002. *Phys. Med. Biol.* **47** 2459-70
[42] Venning A J *et al* 2004 *J. Phys.: Conf. Ser.* **3** 155-8
[43] Hurley C *et al* 2005 *Appl. Appl. Rad. Isotop.* **63** 443-56
[44] Venning A J *et al* 2005 *Phys. Med. Biol.* **50** 3875-88
[45] Venning A J *et al* 2005 *Nuclear Inst. and Methods in Physics Research, A* **555** 396-402
[46] Senden R J *et al* 2006 *Phys. Med. Biol.* **51** 3301-14
[47] Gustavsson H *et al* 2003 *Med. Phys.* **30** 1264-71
[48] Gear J I *et al* 2006 *Phys. Med. Biol.* **51** 3503-16
[49] Lepage M *et al* 2002 *Phys. Med. Biol.* **47** 1881-90
[50] Gore J C *et al* 1996 *Phys. Med. Biol.* **41** 2695-704
[51] Oldham M *et al* 2001 *Med. Phys.* **28** 1436-45
[52] Bosi S G *et al* 2009 *Phys. Med. Biol.* **54** 275-83
[53] Baldock C *et al* 1994 *J. R. Soc. Med.* **87** 806-08
[54] Trapp J V *et al* 2004 *Phys. Med. Biol.* **49** N139-46
[55] Hill B *et al* 2005 *Brit. J. Radiol.* **78** 623-30
[56] Audet C *et al* 2002 *J. Appl. Clin. Med. Phys.* **3** 110-8
[57] Baldock C *et al* 1998 *Phys. Med. Biol.* **43** 3617-27
[58] Rintoul L *et al* 2003 *Appl. Spectroscopy* **57** 51-7
[59] Mather M L *et al* 2002 *Phys. Med. Biol.* **47** 4397-409
[60] Mather M L *et al* 2003. *Ultrasonics* **41** 551-59
[61] Mather M L *et al* 2003. *Phys. Med. Biol.* **48** N269-75
[62] Adamovics J and Maryanski M 2003. *Med. Phys.* **30** 1349
[63] Gorjiara T *et al* 2011. *Med. Phys.* **38** 2265-74
[64] Gorjiara T *et al* 2012 *Med. Phys.* **39** 7071-9
[65] Guo P *et al* 2006 *Med. Phys.* **33** 1338-45
[66] Guo P *et al* 2006 *Med. Phys.* **33** 3962-72
[67] Oldham M *et al* 2008 *Med. Phys.* **35** 2072-80
[68] Gore J C *et al* 1997 *Med. Phys.* **34** 1405-8
[69] Hill B *et al* 2005 *Med. Phys.* **32** 1589-97
[70] Hill B *et al* 2008 *Physica Medica* **24** 149-58
[71] Braun K *et al* 2009 *J. Phys.: Conf. Ser.* **164** 012050
[72] Baldock C *et al* 2010 *Phys. Med. Biol.* **55** R1-63
[73] Baldock C 2006 *J. Phys.: Conf. Ser.* **56** 14-22
[74] Baldock C 2009 *J. Phys.: Conf. Ser.* **164** 012002

Trends in Radiation Dosimetry: preliminary overview of active growth areas, research trends and hot topics from 2011-2015

C Baldock
Faculty of Science, Engineering and Technology, University of Tasmania, Sandy Bay, Tasmania 7005, Australia

Email: clive.Baldock@utas.edu.au

Abstract. The themes and trends of the radiation dosimetry research field were bibliometrically explored by way of co-occurrence term maps using the titles and abstracts text corpora from the Web of Science database for the period from 2011 to 2015. Visualisation of similarities was used by way of the *VOSviewer* visualization tool to generate cluster maps of radiation dosimetry knowledge domains and the associated citation impact of topics within the domains. Heat maps were then generated to assist in the understanding of active growth areas, research trends, and emerging and hot topics.

1. Introduction
Bibliometric indicators such as the h-index are popular metrics used to evaluate researcher performance [1]. Undertaking bibliometric studies of published research papers allows individuals to deduce trends in a specific research field over a specific period and in so doing enable conclusions to be drawn regarding different subfields or research areas [2]. A bibliometric analysis of the radiation dosimetry literature was undertaken from the Web of Science database to investigate the themes and trends of the radiation dosimetry research field to generate cluster maps of knowledge domains and the associated citation impact of topics within the domains. The methodology used so-called term maps to visualize the radiation dosimetry research field. A term map is a two-dimensional representation of a research field in which strongly related terms are located close to each other and less strongly related terms are located further away from each other with the term map providing an overview of the structure of a field. Different areas in a map correspond with different subfields. The term maps were used to assist in the understanding of active growth areas of the radiation dosimetry field and associated research trends and emerging hot topics such as 3-D dosimetry [3, 4]

2. Methods
Titles and abstracts text corpora, corresponding to 5357 publications, was downloaded from the Web of Science database for the period from 2011 to 2015 (figure 1). Title and abstract information for each publication was merged into a single text corpus file for the period under consideration and analysed using the *VOSviewer* visualization tool (www.vosviewer.com) [5-7]. With an emphasis on visualization, the computer program employs a text mining function and associated natural language processing to identify relevant noun phrases in combination with a unified mapping and clustering approach to examine network co-citation data and the co-occurrence of scientific terms. The interactive functionality of the program provides an accessible and hands-on way to explore networks of bibliometric data such as citation counts and/or the co-occurrence relationships among key terms and concepts.

3. Results

3.1 Co-occurrence of terms

Figure 2 shows a co-occurrence cluster map of terms. Each term is represented by a circle, where the diameter of the circle and size of its label represent the frequency of the term, its proximity to another term indicates the degree of relatedness of the two terms, and its colour represents the cluster to which it conceptually belongs. The terms are spatially interrelated in multidimensional space and the figure is limited to a 2-D representation with some relationships between terms not be readily apparent. The map can be seen to contain 5 clusters of co-occurring terms. The red cluster appears more related to dosimetry associated with linear accelerators and treatment planning systems, the green cluster appears more related to dosimetry instrumentation, the yellow cluster appears more related to clinical aspects of dosimetry, the blue cluster appears more related to radiation exposure and dose, and the purple cluster appears more related to radionuclides. Figure 3 shows a co-occurrence term density cluster map.

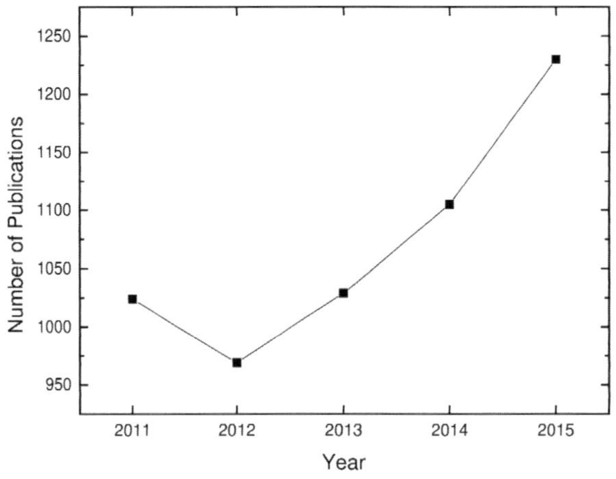

Figure 1. Publications downloaded from the Web of Science database for the period from 2011 to 2015.

3.2 Highly cited terms

Figure 4 is a coloured heat map indicating the relative citation impact of the terms relative to an average citation impact of 1.0 for all publications in the map. Presented data is normalised to take into account the year of publication as older publications have more opportunity for being cited. Red indicates an above average citation impact, blue a below average citation impact and green an average citation impact. Figure 4 clearly indicates that more highly cited publications occur on the right of the figure.

4. Discussion and Conclusions

A number of research topics in the 5-year period have shown significant active growth and are considered to be emerging or hot topics.

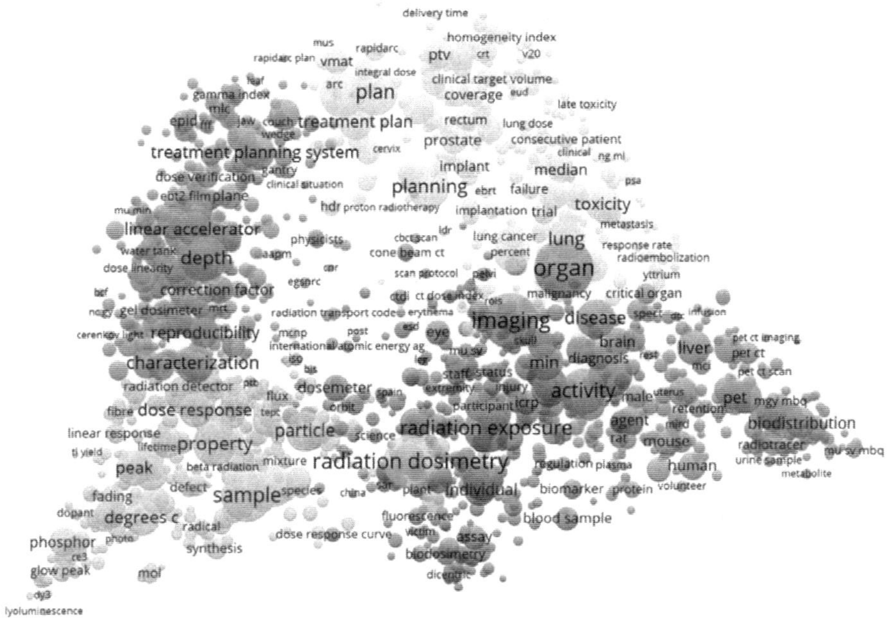

Figure 2. VOSviewer co-occurrence term map.

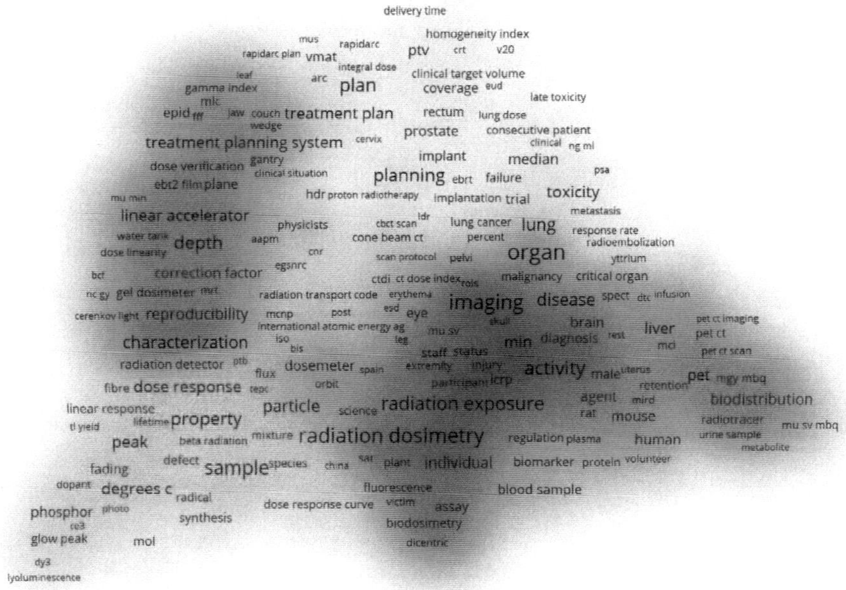

Figure 3. VOSviewer co-occurrence term density cluster map.

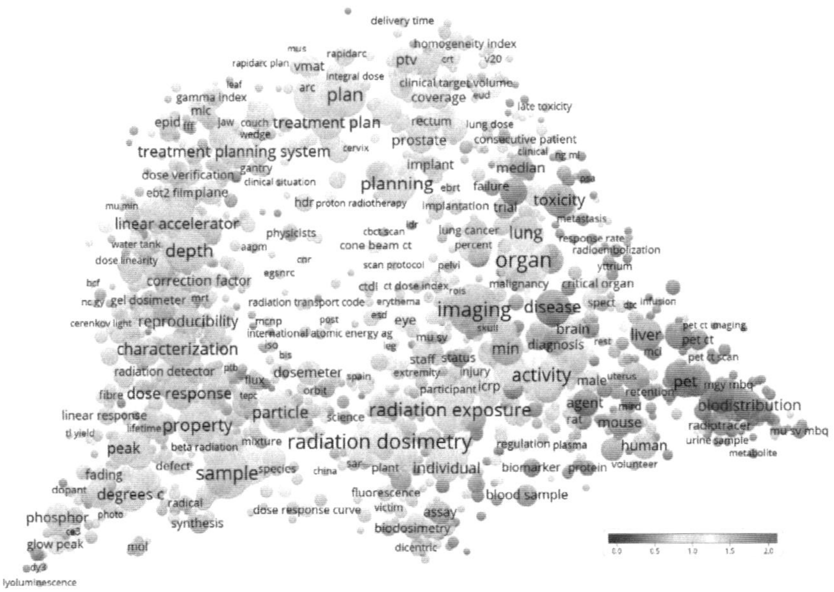

Figure 4. VOSviewer co-occurrence citation impact heat map

The variation in the citation impact across Figure 4 is noticeably with topics on the right of the figure having greater impact in the purple (radionuclides) domain and decreasing through the yellow (clinical aspects of dosimetry) and blue (radiation exposure and dose) domains. There are significant variations in citation practices with, for instance, much larger reference lists in molecular biology than in mathematics resulting in publications in molecular biology on average being cited more frequently than publications in mathematics [8]. This variation in citation practice may account for aspects of that observed in the purple domain.

Further work is continuing beyond this preliminary investigation of active growth areas and research trends to consider the past development of research areas in radiation dosimetry [9-11] with the aim of further identifying current and potential future so-called ‘hot research topics’.

5. References

[1] Baldock C et al 2009 *Med. Phys.* **36** 1043-5
[2] Gorjiara T and Baldock C 2014 *Scientometrics* **100** 121-48
[3] Baldock C *et al* 1998 *Phys. Med. Biol.* **43** 3617-27
[4] Baldock C *et al* 2010 *Phys. Med. Biol.* **55** R1-63
[5] van Eck N J and Waltman L 2010 *Scientometrics* **84** 523-38
[6] van Eck N J et al 2010 *J. American Society for Information Science and Technology* **61** 2405-16
[7] van Eck N J and Waltman L 2011 *ISSI Newsletter* **7** 50-4
[8] van Eck N J *et al* 2013 *PLoS ONE* **8** e62395
[9] Baldock C 2006 *J. Phys.: Conf. Ser.* **56** 14-22
[10] Baldock C 2009 *J. Phys.: Conf. Ser.* **164** 012002
[11] Hill R *et al* 2014 *Phys. Med. Biol.* **59** R183-231

Stereotactic radiosurgery for multiple brain metastases

Anna Lee[1] and Yoshiya (Josh) Yamada[2]
[1]Department of Radiation Oncology, State University of New York (SUNY), Downstate Medical Center, 450 Clarkson Ave, Brooklyn, New York, USA 11203
[2]Department of Radiation Oncology, Memorial Sloan Kettering Cancer Center, 1275 York Ave, New York, New York, USA 10065

E-mail: yamadaj@mskcc.org

Abstract. Whole brain radiation therapy has been the traditional treatment of choice for patients with multiple brain metastases. Although stereotactic radiosurgery is widely accepted for the management to up to 4 brain metastases, its use is still controversial in cases of 5 or more brain metastases. Randomized trials have suggested that stereotactic radiosurgery alone is appropriate in up to 4 metastases without concomitant whole brain radiation. Level 1 evidence also suggests that withholding whole brain radiation may also reduce the impact of radiation on neurocognitive function and also may even offer a survival advantage. A recent analysis of a large multicentre prospective database has suggested that there are no differences in outcomes such as the likelihood of new metastasis or leptomeningeal disease in cases of 2-10 brain metastases, nor in overall survival. Hence in the era of prolonged survival with stage IV cancer, stereotactic radiosurgery is a reasonable alternative to whole brain radiation in order to minimize the impact of treatment upon quality of life without sacrificing overall survival.

1. Introduction

The use of stereotactic radiosurgery for the management of brain metastases is a well accepted paradigm, and multiple randomized trials have demonstrated the efficacy of radiosurgery in providing durable tumour control in better than 80% of cases while the risk of severe complications is extremely rare [1-7]. These trials have limited the number of brain metastases to less than 5 lesions, and traditional treatment of more than 4 brain metastases is typically whole brain radiation (WBRT). WBRT has been recommended in such settings because of the perception that when there are more than 4 metastases, it is likely that many more undetected micrometastases lurk in the brain. Another oft quoted assumption is that with multiple brain metastases, even with stereotactic radiation, the amount of radiation necessary to treat multiple lesions in various quadrants of the brain will amount to whole brain radiation anyways. And in particular for linac users, the time to treat multiple isocenters makes it relatively impractical to treat more than 4 lesions.

2. Why consider radiosurgery

Should stereotactic radiosurgery be considered for patients with greater than five brain metastases? This is a topic of some controversy and there are a number of arguments both in favor and against the use of radiosurgery for multiple metastases in the brain. High dose single fraction radiation therapy is more likely to provide more durable local control as opposed to conventional fractionated radiation treatment [8-10]. Also, focal radiation treatment is likely to spare a significantly greater amount of normal brain parenchyma which should reduce the toxicity of radiotherapy, even for multiple brain metastases. With the advent of modern treatment planning and delivery techniques including image guided single isocenter volumetric arc radiation therapy, in which multiple lesions can be accurately treated with a single isocenter, it is significantly less straining on department resources for treating multiple metastases particularly in the setting of linear accelerator based radiosurgery. In addition, there is data that suggests that stereotactic radiosurgery is cost-effective relative to whole brain radiation. Therefore, there are a several reasons to consider stereotactic radiosurgery for the management of multiple brain metastases.

3. Omitting whole brain radiotherapy

There are a number of studies that suggest that the omission of WBRT and treatment with only stereotactic radiosurgery maybe an appropriate strategy. For example a small randomized trial from Aoyama et al [7] studied patients with 1 to 4 brain metastasis, each less than 3 cm and then randomized them to receive WBRT plus radiosurgery versus radiosurgery alone. In this trial, WBRT patients received 30 Gy in 10 fractions and radiosurgery, adhering closely to RTOG 90-05 guidelines. Radiation dose was reduced by 30% when combined with WBRT in this small randomized trial. There were no significant differences in median survival with the omission of WBRT. However, there was a significantly higher rate of regional brain recurrence at one year and deterioration in mini mental status exam results in patients who underwent WBRT, although this was not a statistically significant finding. Another small series from the University of Pittsburg [11] suggest that the total number of metastases is less relevant than the total volume of tumor to be treated. A multivariate analysis demonstrated that the total treatment tumor volume was significant however the number of metastases was not predictive of overall survival. This study was done with patients with four or more intracranial metastases, thus one large volume metastases is more likely to have a worse outcome than multiple small metastases.

Such data suggest that in the setting of multiple metastases, the addition of WBRT does not offer a significant survival benefit. However, it is likely that radiosurgery does improve the probability of local control. A multi-institutional retrospective study of radiosurgery with or without whole brain radiation was published by Sneed et al [12]. In this large series of patients from 10 different institutions there is no significant survival difference with the admission of upfront whole brain radiation and patience to receive radiosurgery. Median survival times were 14 versus 15.2 months for RPA class one patients, and not significantly different. A meta-analysis of 364 patients from three randomized trials comparing radiosurgery versus whole brain radiation found that for younger patients, omitting whole brain radiation may be reasonable. There was a suggestion toward the reduction in the risk of failure after radiosurgery in patients less than 50 years of age and reduced risk of mortality for those treated with radiosurgery alone for those less than age 70 [13].

In terms of complications from whole brain radiation a randomized trial was reported from Chang et al [14] in which patients who were either RPA class one or two, with 1 versus 2 to 3 brain metastases, and also stratified by histology (melanoma renal cell carcinoma versus other) were randomized to either radiosurgery plus whole brain radiation or radiosurgery alone. The whole brain radiation dose was 30 Gy in 12 fractions, and patients were prospectively evaluated for neurocognitive function. It was found in this small study that the additional whole brain radiation to radiosurgery significantly increased the probability of neurocognitive decline (23% versus 49%, $p = 0.003$). Another interesting finding of this study was that there was actually a survival benefit to delaying WBRT ($p = 0.03$). Patients who underwent whole brain radiation plus radiosurgery had twice the decline in Hopkins Verbal Learning Test and memory compared to patients who were treated with radiosurgery alone. Despite the higher rate of distant brain failure when WBRT was omitted, learning and memory outcomes were still superior for patients who did not receive WBRT. These data suggest that a decline in learning and memory is primarily toxicity from radiation rather than due to brain tumor relapse. The authors suggested that initial treatment for 1 to 3 brain metastases should be radiosurgery alone rather than whole brain radiotherapy in addition to radiosurgery.

Yamamoto et al [3] reported on a large multi-center study of 23 gamma knife centers in Japan for patients with 5 to 10 brain metastasis with a KPS of at least 70 and lesions less than 3 cm in size. Mini mental status exam was utilized for neurologic assessment, then these patients were followed on a regular basis. A total of 1194 patients were treated with radiosurgery alone, with median overall survival of 12 months. A non-inferiority analysis was performed which suggested that although there was a difference when comparing 1 versus more than 1 lesion, when treating 2-4 versus 5-10 lesions, there was no significant difference in overall survival. Similarly, in multivariate analysis, there was a significant difference in overall survival for 1 versus 2 to 10 brain metastases. However, there was no difference in outcomes for 2 to 4 versus 5 to 10 metastasis. The incidence of new lesions with 1 metastasis at initial presentation was found to be 36% at 12 months compared to 54 to 64% for 2 to 4 and 5 to 10 lesions, respectively. At 24 months the likelihood of new metastases in patients who presented with a single metastasis was nearly 50% compared to 65 to 70% for patients with 2 to 4 and 5 to 10 metastases at the time of presentation. There was a significant difference between 1 versus more than 1 but no significant difference between 2 to 4 and 5 to 10 lesions in terms of the likelihood of developing new lesions. Likewise, the risk of leptomeningeal dissemination at 12 months was 7% for patients who presented with a single metastasis versus 8 to 11% if they presented at 2 to 4 or 5 to 10 metastases. And at 24 months, the risk of leptomeningeal disease in a patient who presented with a single lesion was 11% compared to 13% to 21% in patients who had 2 to 4 or 5 to 10 lesions. There was no statistical difference in terms of the likelihood of leptomeningeal disease beyond 1 metastasis. Such data suggest that it may be reasonable to omit whole brain radiation in up to 10 lesions. In terms of mini mental status exam, this study suggests that there was no significant loss of mental status by omitting whole brain radiation despite the presence of up to 10 metastases.

Randomized trials have suggested no overall survival benefit from the addition WBRT radiotherapy. The addition of whole brain radiation to radiosurgery was associated with diminished cognitive functioning and it was also recommended that careful surveillance and judicious use of salvage therapy at the time of brain relapse may be appropriate for patients in order to enjoy a higher quality of life by delaying the initiation of WBRT. Hence, the American Society for Therapeutic Radiation Oncology has suggested that it may be reasonable to omit WBRT upfront for limited brain metastasis [15].

4. Advances in treatment planning and delivery

Particularly among linear accelerator centers, it has been impractical to offer radiosurgery for multiple isocenters. When each lesion is treated with a unique isocenter, radiosurgery to 10 lesions is essentially 10 different courses of treatment, requiring an individual treatment plan for each lesion and isocenter verification for each lesion at the time of treatment. In this case, 10 lesion treatment is extremely resource intensive for both treatment planning and treatment delivery. With the advent of volumetric arc therapy (VMAT), a single isocenter can be utilized to treat multiple lesions, hence a single course of treatment can effectively treat 10 lesions. Although the treatment planning time may be more than for just a single lesion, automated planning tools may be able to effectively minimize the amount of human input necessary. Since treatment will be with a single isocenter, the entire treatment of 10 lesions will require only one isocenter verification, resulting in a much higher throughput for any given linear accelerator. In essence 10 lesions could be treated in nearly the same time required to treat 1 lesion [16].

Although there has been evidence to the contrary, recent data from the University of Alabama in Birmingham have suggested that a single isocenter flattening filter free VMAT is equal in dose distributions to gamma knife radiosurgery for multiple lesions in the brain [16]. Single isocenter multi arc flattening filter free VMAT was found to provide equal plans in terms of plan quality and mean brain dose to that generated with Gamma Knife treatment. A randomized controlled trial is being sponsored by the North American Gamma Knife Consortium for patients with five or more brain metastases and with a KPS > 70. These patients will be randomized to WBRT of 30Gy in 10 fractions versus radiosurgery alone (NCT01731704).

An example of the ability to spare normal tissue is presented in this case of a 52-year-old patient with stage four renal cell carcinoma. This patient was asymptomatic with respect to bone and lung metastases that were controlled on systemic treatment. The patient previously had radiosurgery for a lesion in the right temporal lobe and one in the right parietal lobe. Unfortunately, subsequent scans showed multiple new metastases in the right temporal lobe right, medial posterior frontal lobe, right caudate, left frontal centrum ovale, left occipital temporal junction, left cerebellar, right cerebellar, and left brachium pontis areas of the brain (see figure 1). As figure 2 illustrates, despite having multiple brain metastases, when using radiosurgery, a significant proportion of the brain parenchyma receives less than a single fraction of WBRT (3000cGy/10 fractions).

5. Summary

There is strong rationale treating patients with multiple metastases with radiosurgery alone and omitting upfront WBRT. There appears to be no detrimental loss of overall survival, and level one evidence suggests that there may be a survival advantage in favour of delaying WBRT. There is a significant benefit in terms of reducing the negative effect of radiation on neurocognition when WBRT is avoided. Although there may be differences in survival when comparing patients who receive radiosurgery for 1 metastasis versus more than 1 metastasis, patients with 2-10 brain metastases will enjoy comparable outcomes in terms of overall survival, as well as the likelihood of developing new brain metastases or leptomeningeal disease. Also with advances in treatment planning and delivery, treating multiple brain metastases has become a practical option to consider. Thus, since there are no significant differences in oncological outcomes when more than one lesion is treated, those who are comfortable in offering radiosurgery for 3 to 4 brain metastases should be equally comfortable in offering radiosurgery for up to 10 brain metastases.

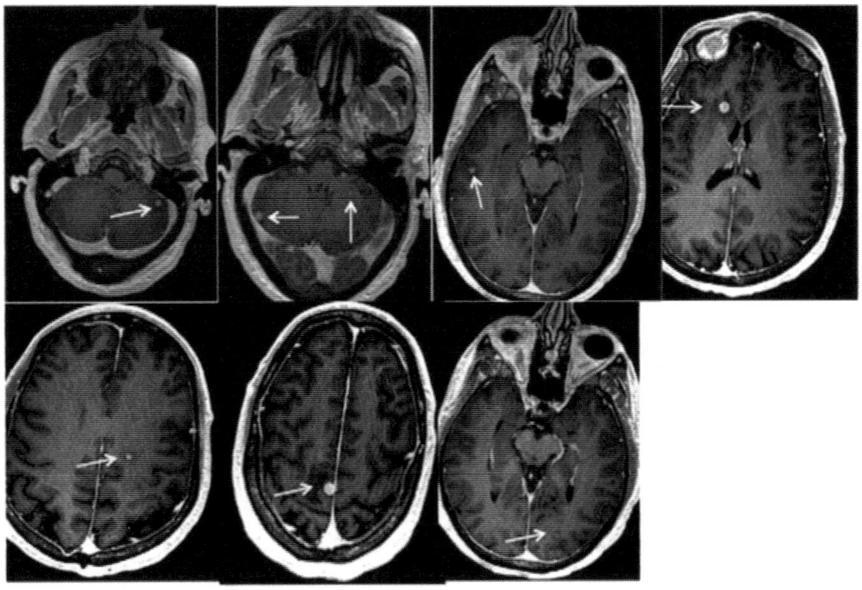

Figure 1. A patient with 8 brain metastases demonstrated on axail T1 weighted post gadolinium contrast enhanced T1 weighted MRI imaging.

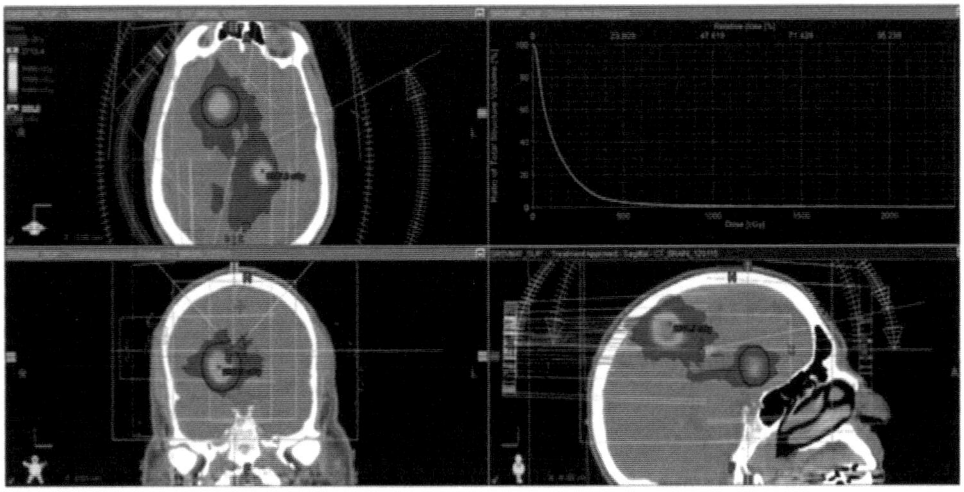

Figure 2. Single isocenter VMAT plan for 8 lesions. The dose color wash is set to 300cGy or greater. Presciption dose was 21 Gy to the 80% isodose line for each lesion.

6. References

[1] Kocher M *et al* 2011 *J. Clin. Oncol.* **29** 134-41.
[2] Alexander E *et al* 1995 *J. Natl. Cancer Inst.* **87** 34-40
[3] Yamamoto M *et al* 2014 *Lancet Oncol.* 15 387-95
[4] Flickinger J C *et al* 1994 *Int. J. Radiat. Oncol. Biol. Phys.* **28** 797-802
[5] Kondziolka D *et al* 1999 *Int. J. Radiat. Oncol. Biol. Phys.* **45** 427-34
[6] Andrews D W *et al* 2004 *Lancet* **363** 1665-72
[7] Aoyama H *et al* 2006 *JAMA* **295** 2483-91
[8] Park S H *et al* 2009 *J. Clin. Neurosci.* **16** 626-9
[9] Suzuki S *et al* 2000 *J. Neurosurg.* **93** 30-1
[10] Kim C H *et al* 2008 *J. Korean Neurosurg.* Soc. **44** 358-63
[11] Bhatnagar A K *et al* 2006 *Int. J. Radiat. Oncol. Biol. Phys.* **64** 898-903
[12] Sneed P K *et al* 2002 *Int. J. Radiat. Oncol. Biol. Phys.* **53** 519-26
[13] Sahgal A *et al* 2015 *Int. J. Radiat. Oncol. Biol. Phys.* **91** 710-7
[14] Chang E L *et al* 2009 *Lancet Oncol.* **10** 1037-44
[15] Tsao M N *et al* 2012 *Pract. Radiat. Oncol.* **2** 210-25
[16] Thomas E M *et al* 2014 *Neurosurgery* **75** 409-18

Overview of Carbon-ion Radiotherapy

Hirohiko Tsujii
Research Center for Charged Particle Therapy, National Institute of Radiological Sciences, Inage-ku Anagawa 4-9-1, Chiba 263-8555, Japan

E-mail: h-tsujii@bj9.so-net.ne.jp

Abstract. The outcome of radiotherapy depends on potential efficiency of accelerators and their related accessories. In charged particle therapy before the 1990s, accelerators that were primarily installed for physics research had been shared, which however had limited flexibility for clinical use. Therapy-dedicated facility was first constructed at Loma Linda University for PBT in 1990 and at NIRS for CIRT in 1993. Currently, there are more than 56 facilities for PBT, 6 for CIRT, and 6 for PBT/CIRT, and even more facilities are under construction or active planning. CIRT has beneficial property for cancer therapy because, as compared with photon therapy, it offers superior dose distributions by exhibiting a Bragg peak in the body and, as compared with PBT, it has higher radiobiological effectiveness. The number of potential candidates for charged particle therapy is estimated to range from 0.018% to 0.035% of all irradiated cancer patients. In CIRT at NIRS, Japan, more than 9,000 patients have been treated with promising results in non-SCC tumors and photon-resistant types of tumors at various sites. It is of note that in CIRT a significant reduction in overall treatment time and fractions has been successfully achieved.

1. Introduction

The primary principle of radiotherapy lies on precise dose localization in the target volume with minimal damage to the surrounding normal tissues. This is particularly so when we notice the historical fact that the photon energy reaching the order of MV contributed significantly to improvement of treatment outcomes. In the 1950s, high-energy accelerators were developed and applied to clinical use, which marked the beginning of modern radiotherapy. In the late 20th century, high-tech radiotherapy such as IMRT and SBRT was developed and charged particle therapy such as proton beam therapy (PBT) and carbon-ion radiotherapy (CIRT) became popular in the world [1, 2].

2. Characteristics of carbon-ion beams

Among several types of ion beam therapy, CIRT has gained a strong interest because they exhibit Bragg peak in the body, enabling delivery of sufficient dose within a target volume and less dose to the surrounding normal tissues. In addition, carbon-ions being heavier than protons provide a higher radiobiological effectiveness (RBE), which increases with depth to reach the maximum at the end of beam's range [3]. This property is extremely advantageous from the therapeutic point of view, as the RBE of carbon ion beams also increases as they advance deeper to the tumor-lying region. The RBE ranges from 2.3 to 3.0 along the SOBP for skin reactions (Fig. 1).

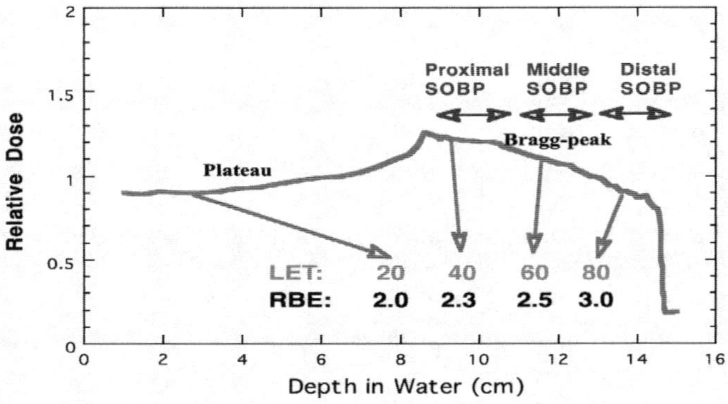

Figure 1. LET and RBE of carbon-ion beams (290MeV, SOBP=60mm).

3. History of particle beam therapy

Historically, the world-first clinical application of particle beams was fast neutron radiotherapy, which started on September 28, 1938 at the Lawrence Berkeley National Laboratory (LBNL), USA. In the late 1970s, this therapy reached its peak in Europe, USA and Japan, and its effectiveness was reported in non-squamous cell tumors and bone/soft-tissue sarcomas. Despite the favourable local control in these tumors, however, fast neutron therapy caused severe late reactions, which eventually led all institutions abandon the therapy. Pi mesons can be generated by collision of high-energy protons or electrons with nucleons in the target. When delivering these elemental particles to the body, they are captured by the medium nucleus at the end of flight path, releasing short-range ions (star production), a mixture of low- to high-LET components. Pi meson radiotherapy was highly expected for its clinical effectiveness and three institutions in the USA, Canada and Switzerland conducted this therapy in the 1980s. However, the clinical outcome was not satisfactory and the clinical research was eventually discontinued. The last institution closed in 1994.

The LBNL pioneered the medical application of proton beams in 1954. Since then, PBT has been increasingly popular in the world. The success of PBT primarily lies on its superior dose distribution so that the definitive dose can be safely delivered with limited toxicities. The same team at LBNL then embarked on helium ions in 1957 and neon ions in 1975 [4]. The Berkeley however terminated all radiotherapy programs in 1992. As if the baton was passed across the Pacific, the NIRS started CIRT in 1994, followed by GSI in 1997 and thereafter by other facilities.

4. Facilities for charged particle therapy

As with the end of 2015, there 56 facilities for PBT and 12 facilities for CIRT including 6 facilities for CIRT alone and 6 facilities for both CIRT and PBT [2]. In addition, 4 new facilities are constructing carbon therapy centers (Japan 2, China 1, Korea 1).

CIRT was initiated in 1994 at NIRS using the world-first accelerator complex (HIMAC: Heavy Ion Medical Accelerator in Chiba) dedicated to cancer therapy. Following HIMAC, Gesellschaft für Schwerionenforschung (GSI) in Germany started CIRT in 1997, which was succeeded in 2009 by the proton/carbon center, the Heidelberg Ion-Beam Therapy Center (HIT). Marburg University also started proton/carbon therapy in 2015. Hyogo Ion Beam Medical Center (HIBMC) in Japan was the first proton/carbon center established in 2001. The Institute of Modern Physics (IMP), China, started clinical trials in 2006. The IMP constructed a carbon center in Wuwei and is also constructing a new center near the yellow river. Based on technological development at NIRS, a downsized carbon facilities were realized at Gunma University Heavy Ion Medical Center (GHMC), SAGA Heavy Ion Medical Accelerator in Tosu (SAGA HIMAT) and Kanagawa Cancer Center (iROCK). Each of them started CIRT in 2010, 2013 and 2015, respectively. At the National Centre of Oncological Hadrontherapy

(CNAO), Italy, the accelerator complex was completed for proton/carbon treatment in 2011. In the summer of 2014, Shanghai Proton Heavy Ion Center (SPHIC), China, started clinical application of proton/carbon beams. MedAustron in Austria will start clinical study in 2016. Two institutions are under construction for CIRT in Yamagata and Osaka in Japan.

Regarding the status of charged particle therapy in the United States, there are around 23 facilities for PBT but none for CIRT. However, the NCI recently issued two calls for P20 exploratory grant applications for Planning for a National Center for Particle Beam Radiation Therapy Research. In this framework, the NCI awards will support planning for establishment of a Center for charged particle therapy as a national research resource, as well as the DOE awards will address development of improved hardware that could shrink the size, increase the maneuverability, and considerably reduce the steep costs of charged particle therapy equipments. In the spring of 2015, the NCI awarded University of Texas Southwestern Medical Center (UTSW) in Dallas and Particle Therapy Research Group of UCSF in San Francisco a planning grant to develop research proposal for the center. They have decided to establish a heavy ion therapy center and started collaboration with NIRS, CNAO and HIT. The UTSW has already started a UTSW Master Plan for a PBT/CIRT center to be built in their campus. UCSF organized the North American Particle Therapy Alliance (NAPTA) to start the feasibility study for charged particle therapy including CIRT.

5. Potential indications for charged particle radiotherapy

In treatment of histologically radio-resistant and locally advanced tumors, treatment outcome has been improved with high-LET radiations like carbon ions. In CIRT the very sharp lateral fall-off is particularly beneficial, as critical organs next to the tumor are effectively spared from excessive radiation dose. According to PTCOG statistics, a total of 154,097 patients were treated with charged particle therapy in the world from 1954 to December 2015, of which 131,134(85.1%) were treated with protons and 193,76(12.6%) with carbon ions. The selection of patients for CIRT is primarily based on the following 4 clinical criteria: 1) loco-regional tumor growth, 2) tumor with no or low tendency to metastasize, 3) tumor that is highly resistant to conventional radiotherapy, and 4) either cure rates are low or side effects with standard therapy are unacceptably high.

Table 1. How many patients need particle beam therapy?

Author	Country	Population (a)	Annual number of patients treated with RT (b)	Patients eligible for charge particle therapy per year		
				% of pats in RT pats (c)	No. pats bxc/100 (d)	% d/ax100
Bengt (2005)	Sweden	9,593,000	16,000	14~15%	2,200~2,500	0.024%
Orecchia (1998)	Italy	59,830,000	-	-	11,000	0.018%
Baron (2004)	France	66,030,000	158,620	14.5%	23,000	0.035%
Mayer (2004)	Austria	8,474,000	15,132	13.5%	2,043	0.024%
Mella (2013)	Norway	5,100,000	16,000~19,000	7~8%	1,100~1,500	0.025%
Tsujii (2016)	Japan	127,300,000	201,000	14.9%	30,000	0.024%

There have been several studies investigating the potential number of cancer patients that could benefit from charged particle therapy. Using domestic and overseas cancer statistics for incidence and trend of treatment, we have made a strict estimation regarding the number of potential candidates for charged particle therapy, which ranged from 13% to 15% of all irradiated cancer patients. In Japan, the total number of the patients treated with radiotherapy was 201,000 in 2009, of which approximately 30,000 patients or 14.9% was estimated to benefit from charged particle therapy. The absolute number of the patients suitable for charged particle therapy should vary according to the total number of cancer patients and is estimated to range from 0.018% to 0.035% of the population in each country (Table 1).

6. Clinical results of carbon-ion radiotherapy

Since 1994, CIRT has been conducted at the National Institute of Radiological Sciences (NIRS; Chiba, Japan), where more than 9,000 patients with various types of tumors have been treated. In other CIRT facilities in the world, the number of the patients has rapidly increased totalling more than 15,000 patients. Latest experiences on CIRT has indicated improved results in: non-SCC type of head and neck cancer; bone/soft tissue sarcoma arising from the skull base, head-and-neck, spine and pelvis; locally advanced large tumors at various sites; intermediate to high-risk prostate cancer; pelvic recurrence of rectal cancer; and locally advanced unresectable pancreatic cancer (LAUPC) [5]. Among them, the results of CIRT for LAUPC were promising and caused worldwide interest in conducting a comparative study on this tumor. Accordingly, even a randomized Phase 3 trial on "Carbon Ion versus Photon Therapy for Pancreatic Cancer" has been proposed [6].

Experiments with fast neutron beams as well as carbon-ion beams have demonstrated that increasing the dose per fraction tended to lower the RBE for both tumor and normal tissues, but the RBE for the tumor did not decrease as rapidly as the RBE for normal tissues [7]. Accordingly, in CIRT at NIRS a significant reduction in overall treatment time and fractions has been obtained with minor toxicities, such as single-fraction RT for early-stage lung cancer, single or two-fraction RT for liver cancer and 12-fraction RT for prostate cancer. Also for other tumors, 16 or even smaller fractions have been sufficient. This means that CIRT can be performed more efficiently than PBT, permitting treatment for a larger number of patients than is possible with PBT over the same period of time. Further reducing the number of fractions would increase cost-utility and patient throughput, thereby facilitating the use of CIRT for more patients and in more centers.

7. References

[1] Schulz-Ertner D and Tsujii H 2007 *J. Clin. Oncol.* **25** 953
[2] Jerman M 2016 *PTCOG* (http://www.ptcog.ch)
[3] Tsujii H *et al* 2014 *Carbon-Ion Radiotherapy; Principles, Practices and Treatment Planning* (Japan; Springer-Verlag)
[4] Castro J 1995 *Radiat. Environ. Biophys.* **34** 45
[5] Tsujii H and Kamada T 2012 *Japan. J. Clin. Oncol.* **42** 670
[6] Durante M *et al* 2015 *Lancet Oncol.* **16** e93-100.
[7] Ando K *et al* 2005 *J. Radiat. Res.* **46** 51